ELEMENTS OF OPTICAL COMMUNICATION AND OPTO ELECTRONICS

ELEMENTS OF OPTICAL COMMUNICATION AND OPTO ELECTRONICS

By

ANIL KUMAR SHUKLA

M. Sc. (Electronics), M. Tech. (Ph.D.)

Asst. Professor,

AMITY Institute of Telecom Technology and Management,

AMITY University, Noida (U.P.)

UNIVERSITY SCIENCE PRESS

(An Imprint of Laxmi Publications Pvt. Ltd.)

BANGALORE • CHENNAI • COCHIN • GUWAHATI • HYDERABAD
JALANDHAR • KOLKATA • LUCKNOW • MUMBAI • PATNA
RANCHI • NEW DELHI

Published by :
UNIVERSITY SCIENCE PRESS
(An Imprint of Laxmi Publications Pvt. Ltd.)
113, Golden House, Daryaganj,
New Delhi-110002

Phone : 011-43 53 25 00
Fax : 011-43 53 25 28

www.laxmipublications.com
info@laxmipublications.com

First Edition : 2012

OFFICES

© **Bangalore**	080-26 75 69 30	© **Chennai**	044-24 34 47 26
© **Cochin**	0484-237 70 04, 405 13 03	© **Guwahati**	0361-251 36 69, 251 38 81
© **Hyderabad**	040-24 65 23 33	© **Jalandhar**	0181-222 12 72
© **Kolkata**	033-22 27 43 84	© **Lucknow**	0522-220 99 16
© **Mumbai**	022-24 91 54 15, 24 92 78 69	© **Patna**	0612-230 00 97
© **Ranchi**	0651-220 44 64		

UOC-9629-125-OPTI COMM & OPTO ELECT-SHU **C—**

Typeset at : ABRO Enterprises, Delhi. Printed at : Repro Knowledgecast Limited, Thane

PREFACE

Due to tremendous increase in bandwidth capability brings a convergence of voice, video and data over a single fiber-optic channel. An all-optical network will be part of a complex combination of networks including wireless and wireline technologies developed throughout the world. The number and complexity of fiber optic devices and systems has increased dramatically, and the skills and knowledge required to work on these systems has expanded as well. The book is divided into eight chapters as given as under:

Chapter 1 gives background on fiber optics.

Chapter 2 discusses the fiber optic concepts.

Chapter 3 introduces about optical fibers and cables.

Chapter 4 deals with optical splices, connectors and couplers.

Chapter 5 discusses fiber optic measurement techniques.

Chapter 6 gives optical sources and fiber optic transmitters.

Chapter 7 introduces optical detectors and fiber optic receivers.

Chapter 8 introduces fiber optic links.

The book will be useful for engineering students to understand the basic concepts of optical fiber, opto electronics and optical fiber communication without the use of rigorous mathematical explanation. It will also serve for those students who are preparing for various competitive examinations and appearing for interview in various telecom companies. In addition to that the book will also cater for professionals involved with fiber optics, service engineers and technicians.

—Author

ACKNOWLEDGEMENT

First of all I would like to thank almighty GOD, who motivated me throughout my work. I would like to express my sincere thanks and gratitude to Lt. Gen. P.D. Bhargava, Director General, Amity Institute of Telecom technology and Management, Amity University Noida for his constant encouragement and suggestions.

I also feel privileged to thank Dr. Ashok K. Chauhan, Founder president, RBEF for providing environment for this work. I also thank Dr. R.K. Tiwari, Reader, department of physics and electronics, Dr. R.M.L., Avadh University, Faizabad for his regular suggestions and help.

My special thanks to Brig. (retd.) B.K. Kohli, Dean (Academics), Amity Institute of Telecom Technology and Management for his continuous support and motivation.

I would also like to acknowledge all my colleagues and friends for their help and suggestions during this project.

I am grateful to my family members especially my wife Anshu for inspiring me for this project.

CONTENTS

1 BACKGROUND ON FIBER OPTICS

Learning Objectives

Learning objectives are stated at the beginning of each chapter. These learning objectives serve as a preview of the information you are expected to learn in the chapter. The comprehensive check questions are based on the objectives. By successfully completing the NRTC, you indicate that you have met the objectives and have learned the information. The learning objectives are listed below.

Upon completing this chapter, you should be able to do the following:

- Describe the term **fiber optics.**
- List the parts of a fiber optic data link.
- Understand the function of each fiber optic data link part.
- Outline the progress made in the history of fiber optic technology.
- Describe the trade-offs in fiber properties and component selection in the design of fiber optic systems.
- List the advantages and the disadvantages of fiber optic systems compared.

DEFINITION OF FIBER OPTICS

In the past 30 years, researchers have developed a new technology that offers greater data rates over longer distances at costs lower than copper wire systems. This new technology is **fiber optics.** Fiber optics uses light to send information (data). More formally, **fiber optics** is the branch of optical technology concerned with the transmission of radiant power (light energy) through fibers.

FIBER OPTIC DATA LINKS

A fiber optic data link sends input data through fiber optic components and provides this data as output information. It has the following three **basic functions:**

- To convert an electrical input signal to an optical signal.
- To send the optical signal over an optical fiber.
- To convert the optical signal back to an electrical signal.

A fiber optic data link consists of three parts—**transmitter, optical fiber** and **receiver.** Figure 1.1 is an illustration of a fiber optic data-link connection. The transmitter, optical fiber, and receiver perform the basic functions of the fiber optic data link. Each part of the data link is responsible for the successful transfer of the data signal.

A fiber optic data link needs a transmitter that can effectively convert an electrical input signal to an optical signal and launch the data-containing light down the optical fiber. A fiber optic data link also needs a receiver that can effectively transform this optical signal back into its original form. This means that the electrical signal provided as data output should exactly match the electrical signal provided as data input.

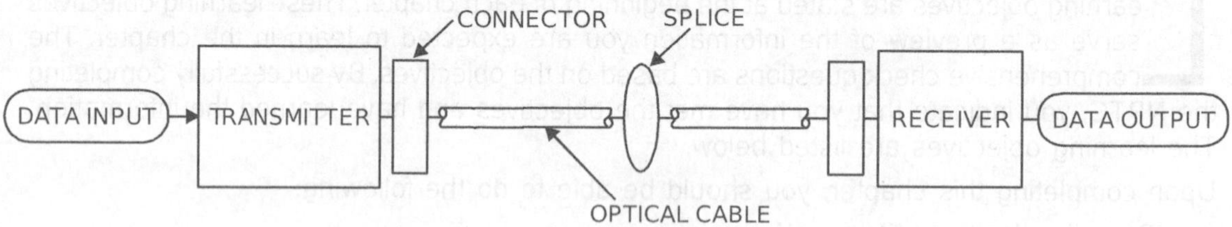

Fig. 1.1. *Parts of a fiber optic data link.*

The transmitter converts the input signal to an optical signal suitable for transmission. The transmitter consists of two parts, an interface circuit and a source drive circuit. The transmitter's drive circuit converts the electrical signals to an optical signal. It does this by varying the current flow through the light source. The two types of optical sources are light-emitting diodes (LEDs) and laser diodes.

The optical source launches the optical signal into the fiber. The optical signal will become progressively weakened and distorted because of scattering, absorption, and dispersion mechanisms in the fiber waveguides. Chapter 2 discusses the fiber mechanisms of scattering, absorption, and dispersion. Chapter 3 discusses the types of optical fibers and cables.

The receiver converts the optical signal exiting the fiber back into an electrical signal. The receiver consists of two parts, the optical detector and the signal-conditioning circuits. An optical detector detects the optical signal. The signal-conditioning circuit conditions the detector output so that the receiver output matches the original input to the transmitter. The receiver should amplify and process the optical signal without introducing noise or signal distortion. Noise is any disturbance that obscures or reduces the quality of the signal. Noise effects and limitations of the signal-conditioning circuits cause the distortion of the receiver's electrical output signal.

An optical detector can be either a semiconductor positive-intrinsic-negative (PIN) diode or an avalanche photodiode (APD).

A PIN diode changes its electrical conductivity according to the intensity and wavelength of light. The PIN diode consists of an intrinsic region between *p*-type and *n*-type semiconductor material. Chapter 6 provides further explanation of optical sources. Chapter 7 provides further explanation of optical detectors.

A fiber optic data link also includes passive components other than an optical fiber. Figure 1.1 does not show the optical connections used to complete the construction of the fiber optic data link. Passive components used to make fiber connections affect the performance of the data link. These components can also prevent the link from operating. Fiber optic components used to make the optical connections include optical splices, connectors and couplers. Chapter 4 outlines the types of optical splices, connectors and couplers and their connection properties that affect system performance.

Proof of link performance is an integral part of the design, fabrication and installation of any fiber optic system. Various measurement techniques are used to test individual parts of a data link. Each data link part is tested to be sure the link is operating properly. Chapter 5 discusses the laboratory and field measurements used to measure link performance.

HISTORY OF FIBER OPTIC TECHNOLOGY

People have used light to transmit information for hundreds of years. However, it was not until the 1960s, with the invention of the laser that widespread interest in optical (light) systems for data communications began. The invention of the laser prompted researchers to study the potential of fiber optics for data communications, sensing, and other applications. Laser systems could send a much larger amount of data than telephone, microwave, and other electrical systems. The first experiment with the laser involved letting the laser beam transmit freely through the air. Researchers also conducted experiments letting the laser beam transmit through different types of waveguides. Glass fibers, gas-filled pipes, and tubes with focusing lenses are examples of optical waveguides. Glass fibers soon became the preferred medium for fiber optic research.

Initially, the very large losses in the optical fibers prevented coaxial cables from being replaced. **Loss** is the decrease in the amount of light reaching the end of the fiber. Early fibers had losses around 1,000 dB/km making them impractical for communications use. In 1969, several scientists concluded that impurities in the fiber material caused the signal loss in optical fibers. The basic fiber material did not prevent the light signal from reaching the end of the fiber. These researchers believed it was possible to reduce the losses in optical fibers by removing the impurities. By removing the impurities, construction of low-loss optical fibers was possible.

There are two basic types of optical fibers, **multimode fibers** and **single mode fibers.** Chapter 2 discusses the differences between the fiber types.

In 1970, Corning Glass Works made a multimode fiber with losses under 20 dB/km.

This same company, in 1972, made a high silica-core multimode optical fiber with 4 dB/km minimum attenuation (loss). Currently, multimode fibers can have losses as low as 0.5 dB/km at wavelengths around 1300 nm. Single mode fibers are available with losses lower than 0.25 dB/km at wavelengths around 1500 nm.

Developments in semiconductor technology, which provided the necessary light sources and detectors, furthered the development of fiber optics. Conventional light sources, such as lamps or lasers, were not easily used in fiber optic systems. These light sources tended to be too large and required lens systems to launch light into the fiber. In 1971, Bell Laboratories developed a small area light-emitting diode (LED). This light source was suitable for low-loss coupling to optical fibers. Researchers could then perform source-to-fiber jointing easily and repeatedly. Early semiconductor sources had operating lifetimes of only a few hours. However, by 1973, projected lifetimes of lasers advanced from a few hours to greater than 1,000 hours. By 1977, projected lifetimes of lasers advanced to greater than 7,000 hours. By 1979, these devices were available with projected lifetimes of more than 1,00,000 hours.

In addition, researchers also continued to develop new fiber optic parts. The types of new parts developed included low-loss fibers and fiber cables, splices, and connectors. These parts permitted demonstration and research on complete fiber optic systems.

Advances in fiber optics have permitted the introduction of fiber optics into present applications. These applications are mostly in the telephone long-haul systems, but are growing to include cable television, computer networks, video systems, and data links. Research should increase system performance and provide solutions to existing problems in conventional applications. The impressive results from early research show there are many advantages offered by fiber optic systems.

FIBER OPTIC SYSTEMS

System design has centered on long-haul communications and the subscriber-loop plant. The subscriber-loop plant is the part of a system that connects a subscriber to the nearest switching center. Cable television is an example. Limited work has also been done on short-distance applications and some military systems.

Initially, central office trunking required multimode optical fibers with moderate to good performance. Fiber performance depends on the amount of loss and signal distortion introduced by the fiber when it is operating at a specific wavelength. Long-haul systems require single mode optical fibers with very high performance. Single mode fibers tend to have lower loss and produce less signal distortion.

In contrast, short-distance and military systems tend to use only multimode technology. Examples of short-distance systems include process control and local area networks (LANs). Short-distance and military systems have many connections. The larger fiber core and higher fiber numerical aperture (NA) of multimode fibers reduce losses at these connections. Chapter 4 explains fiber connection properties in more detail. Chapter 2 provides more detail on multimode and single mode fibers.

In military and subscriber-loop applications, system design and parts selection are related. Designers consider **trade-offs** in the following areas:

- Fiber properties
- Types of connections
- Optical sources
- Detector types

Designers develop systems to meet stringent working requirements, while trying to maintain economic performance. It is quite difficult to identify a standard system design approach. This module identifies the types of components chosen by the Navy for shipboard applications.

Future system design improvements depend on continued research. Researchers expect fiber optic product improvements to upgrade performance and lower costs for short-distance applications. Future systems center on broadband services that will allow transmission of voice, video, and data. Services will include television, data retrieval, video word processing, electronic mail, banking, and shopping.

ADVANTAGES AND DISADVANTAGES OF FIBER OPTICS

Fiber optic systems have many attractive features that are superior to electrical systems. These include improved system performance, immunity to electrical noise, signal security, and improved safety and electrical isolation.

Other advantages include reduced size and weight, environmental protection, and overall system economy. The main advantages of fiber optic systems are given as under:

- System performance
- Greatly increased bandwidth and capacity
- Lower signal attenuation (loss)
- Immunity to electrical noise
- Immune to noise (electromagnetic interference [EMI] and radio-frequency interference [RFI])
- No crosstalk
- Lower bit error rates
- Signal security
- Difficult to tap
- Non-conductive (does not radiate signals) Electrical Isolation
- No common ground required
- Freedom from short circuit and sparks
- Size and weight
- Reduced size and weight cables
- Environmental protection
- Resistant to radiation and corrosion
- Resistant to temperature variations

- Improved ruggedness and flexibility
- Less restrictive in harsh environments
- Overall system economy
- Low per-channel cost.

Limitations of Fiber Optic Systems

Despite the many advantages of fiber optic systems, there are some disadvantages.

Because of the relative newness of the technology, fiber optic components are expensive. Fiber optic transmitters and receivers are still relatively expensive compared to electrical interfaces. The lack of standardization in the industry has also limited the acceptance of fiber optics. Many industries are more comfortable with the use of electrical systems and are reluctant to switch to fiber optics. However, industry researchers are eliminating these disadvantages.

Standards committees are addressing fiber optic part and test standardization.

The cost to install fiber optic systems is falling because of an increase in the use of fiber optic technology. Published articles, conferences, and lectures on fiber optics have begun to educate managers and technicians. As the technology matures, the use of fiber optics will increase because of its many advantages over electrical systems.

SUMMARY

Now that you have completed this chapter, let's review some of the new terms, concepts, and ideas you have learned. You should have a thorough understanding of these principles before advancing to Chapter 2.

FIBER OPTICS is the branch of optical technology concerned with the transmission of radiant power (light energy) through fibers.

A FIBER OPTIC DATA LINK has three basic functions: to convert an electrical input signal to an optical signal, to send the optical signal over an optical fiber, and to convert the optical signal back to an electrical signal. It consists of three parts: transmitter, optical fiber and receiver.

The TRANSMITTER consists of two parts, an interface circuit and a source drive circuit. The transmitter converts the electrical input signal to an optical signal by varying the current flow through the light source.

The RECEIVER consists of two parts, the optical detector and signal conditioning circuits. The receiver converts the optical signal exiting the fiber back into the original form of the electrical input signal.

SCATTERING, ABSORPTION and DISPERSION MECHANISMS in the fiber waveguides cause the optical signal launched into the fiber to become weakened and distorted.

NOISE is any disturbance that obscures or reduces the quality of the signal.

SIGNAL LOSS is the decrease in the amount of light reaching the end of the fiber. Impurities in the fiber material cause the signal loss in optical fibers. By removing these impurities, construction of low-loss optical fibers was possible.

The TWO BASIC TYPES OF OPTICAL FIBERS are multimode fibers and single mode fibers.

A LOW-LOSS MULTIMODE OPTICAL FIBER was developed in 1970.

A SMALL AREA LIGHT-EMITTING DIODE (LED) was developed in 1971. This light source was suitable for low-loss coupling to optical fibers.

FIBER OPTIC SYSTEM DESIGN has centered on long-haul communications and the subscriber-loop plant. Limited work has also been done on short-distance applications and some military systems.

FIBER PERFORMANCE depends on the amount of loss and signal distortion introduced by the fiber when it is operating at a specific wavelength. Single mode fibers tend to have lower loss and produce less distortion than multimode fibers.

In MILITARY and SUBSCRIBER-LOOP APPLICATIONS, **system designers consider trade-offs in the following areas: fiber properties, types of connections, optical sources** and **detector types.**

The **ADVANTAGES** of fiber optic systems include improved system performance, immunity to electrical noise, signal security and electrical isolation. Advantages also include reduced size and weight, environmental protection and overall system economy.

The **DISADVANTAGES** of fiber optic systems include problems with the relative newness of the technology, the relatively expensive cost and the lack of component and system standardization. However, these disadvantages are already being eliminated because of increased use and acceptance of fiber optic technology.

Electrical noise, signal security, electrical isolation, reduced size and weight, environmental protection and overall system economy.

REVIEW QUESTIONS AND ANSWERS

Q. 1. *Define fiber optics.*

Ans. Fiber optics is the branch of optical technology concerned with the transmission of radiant power (light energy) through fibers.

Q. 2. *Describe the basic functions of a fiber optic data link.*

Ans. The basic functions of a fiber optic data link are to convert an electrical input signal to an optical signal, send the optical signal over an optical fiber, and convert the optical signal back to an electrical signal.

Q. 3. *List the three parts of a fiber optic data link.*

Ans. Transmitter, optical fiber and receiver.

Q. 4. *What mechanisms in the fiber waveguides weaken and distort the optical signal?*

Ans. Scattering, absorption and dispersion.

Q. 5. *What effect does noise has on the fiber optic signal?*

Ans. Noise obscures or reduces the quality of the signal.

Q. 6. *Define loss.*

Ans. Loss is the decrease in the amount of light reaching the end of the fiber.

Q. 7. *In 1969, what did several scientists conclude about optical fiber loss?*

Ans. Impurities in the fiber material caused the signal loss in optical fibers. The basic fiber material did not prevent the light signal from reaching the end of the fiber.

Q. 8. *How can loss be reduced during construction (or fabrication) of optical fibers?*

Ans. By removing the impurities from optical fiber.

Q. 9. *What are the two basic types of optical fibers?*

Ans. Multimode and single mode fibers.

Q. 10. *Which type of optical fiber (multimode or single mode) tends to have lower loss and produces less signal distortion?*

Ans. Single mode fiber.

Q. 11. *What optical fiber properties reduce connection loss in short-distance systems?*

Ans. Larger fiber core and higher fiber numerical aperture (NA).

Q. 12. *In fiber optic systems, designers consider what trade-offs?*

Ans. Trade-offs in fiber properties, types of connections, optical sources, and detector types in military and subscriber-loop applications.

Q. 13. *List seven advantages of fiber optics over electrical systems.*

Ans. Advantages of fiber optics are improved system performance, immunity to electrical noise, signal security, electrical isolation, reduced size and weight, environmental protection and overall system economy.

■■■

2 FIBER OPTIC CONCEPTS

Learning Objectives

Upon completion of this chapter, you should be able to do the following:

- Understand the nature of light propagation.
- Discuss the electromagnetic theory of light.
- Describe the properties of light reflection, refraction, diffusion, and absorption.
- Explain how optical fibers transmit light. Identify the basic optical fiber material properties.
- Describe the ray and mode theories of light propagation along an optical fiber.
- State the difference between multimode and single mode optical fibers.
- Explain how optical fibers attenuate and distort light signals as they travel along the optical fiber.
- Understand the processes of light attenuation and dispersion.

FIBER OPTIC LIGHT TRANSMISSION

Fiber optics deals with the transmission of light energy through transparent fibers. How an optical fiber guides light depends on the nature of the light and the structure of the optical fiber. A light wave is a form of energy that is moved by wave motion. Wave motion can be defined as a recurring disturbance advancing through space with or without the use of a physical medium. In fiber optics, wave motion is the movement of light energy through an optical fiber.

PROPAGATION OF LIGHT

The exact nature of light is not fully understood, although people have been studying the subject for many centuries. In the 1700s and before, experiments seemed to indicate that light was composed of particles. In the early 1800s, a physicist Thomas Young showed that light exhibited wave characteristics.

Further experiments by other physicists culminated in James Clerk (pronounced Clark) Maxwell collecting the four fundamental equations that completely describe the behaviour of the electromagnetic fields. James Maxwell deduced that light was simply a component of the electromagnetic spectrum. This seems to firmly establish that light is a wave. Yet, in the early 1900s, the interaction of light with semiconductor materials, called the **photoelectric effect,** could not be explained with electromagnetic-wave theory.

The advent of quantum physics successfully explained the photoelectric effect in terms of fundamental particles of energy called **quanta.** Quanta are known as **photons** when referring to light energy.

Today, when studying light that consists of many photons, as in propagation, that light behaves as a continuum—an electromagnetic wave. On the other hand, when studying the interaction of light with semiconductors, as in sources and detectors, the quantum physics approach is taken. The wave versus particle dilemma can be addressed in a more formal way, but that is beyond the scope of this text. It suffices to say that much has been reconciled between the two using quantum physics. In this manual, we use both the electromagnetic wave and photon concepts, each in the places where it best matches the phenomenon we are studying.

The electromagnetic energy of light is a form of electromagnetic radiation.

Light and similar forms of radiation are made up of moving electric and magnetic forces. A simple example of motion similar to these radiation waves can be made by dropping a pebble into a pool of water. In this example, the water is not actually being moved by the outward motion of the wave, but rather by the up-and-down motion of the water. The up-and-down motion is transverse, or at right angles, to the outward motion of the waves. This type of wave motion is called **transverse-wave motion.** The transverse waves spread out in expanding circles until they reach the edge of the pool, in much the same manner as the transverse waves of light spread from the sun. However, the waves in the pool are very slow and clumsy in comparison with light, which travels approximately 186,000 miles per second.

Light radiates from its source in all directions until it is absorbed or diverted by some substance (Fig. 2.1). The lines drawn from the light source (a light bulb in this instance) to any point on one of the transverse waves indicate the direction that the wavefronts are moving. These lines are called **light rays.**

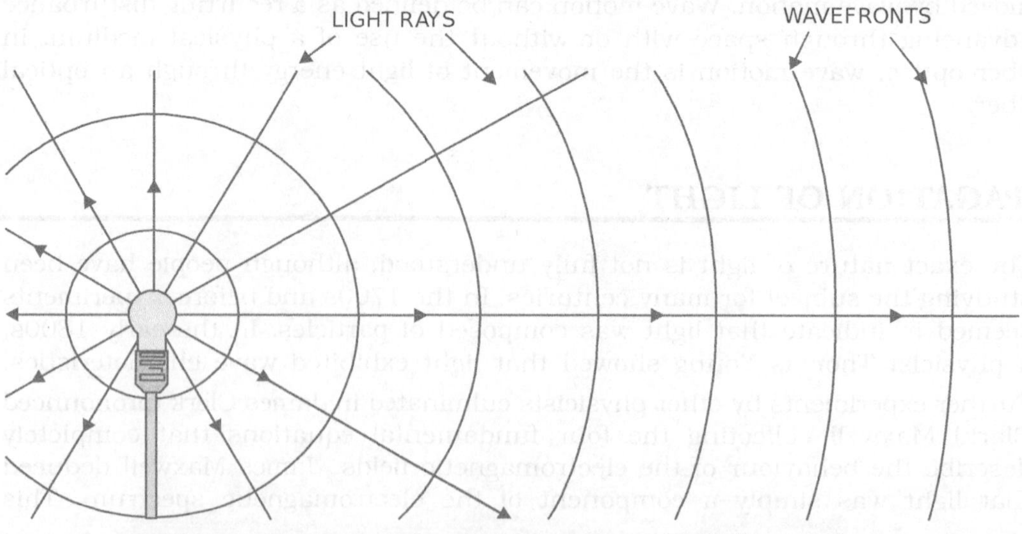

Fig. 2.1. *Light rays and wavefronts from a nearby light source*

Although single rays of light typically do not exist, light rays shown in illustrations are a convenient method used to show the direction in which light is travelling at any point. A ray of light can be illustrated as a straight line.

To fully understand the concept of wave motion, refer to NEETS Module 10—*Introduction to wave propagation, transmission lines* and *antennas.* Before we introduce the subject of light transmission through optical fibers, you must first understand the nature of light and the properties of light waves.

PROPERTIES OF LIGHT

When light waves, which travel in straight lines, encounter any substance, they are reflected, absorbed, transmitted, or refracted. This is illustrated in Fig. 2.2. Those substances that transmit almost all the light waves falling upon them are said to be **transparent.** A transparent substance is one through which you can see clearly.

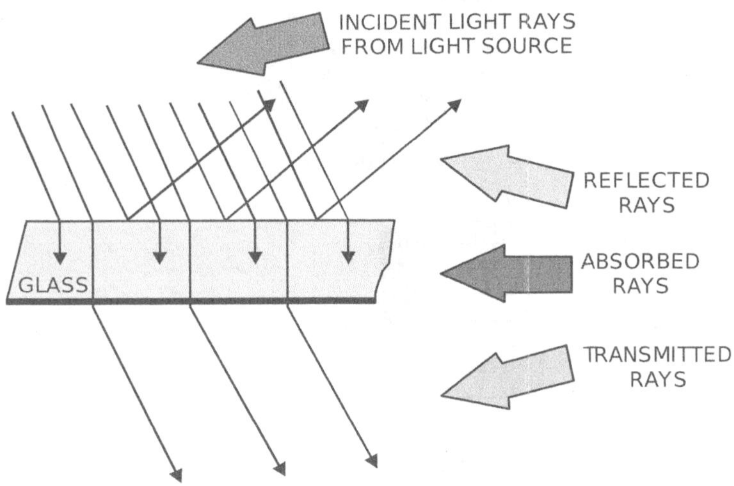

Fig. 2.2. *Light waves reflected, absorbed and transmitted*

Clear glass is transparent because it transmits light rays without diffusing them (view A of Fig. 2.3). There is no substance known that is perfectly transparent, but many substances are nearly so. Substances through which some light rays can pass, but through which objects cannot be seen clearly because the rays are diffused, are called **translucent** (view B of Fig. 2.3). The frosted glass of a light bulb and a piece of oiled paper are examples of translucent materials. Those substances that are unable to transmit any light rays are called **opaque** (view C of Fig. 2.3). Opaque substances either reflect or absorb all the light rays that fall upon them.

All substances that are not light sources are visible only because they reflect all or some part of the light reaching them from some luminous source.

Examples of luminous sources include the sun, a gas flame, and an electric light filament, because they are sources of light energy. If light is neither transmitted nor reflected, it is absorbed or taken up by the medium. When light strikes a substance, some absorption and some reflection always take place. No substance completely transmits, reflects, or absorbs all the light rays that reach its surface.

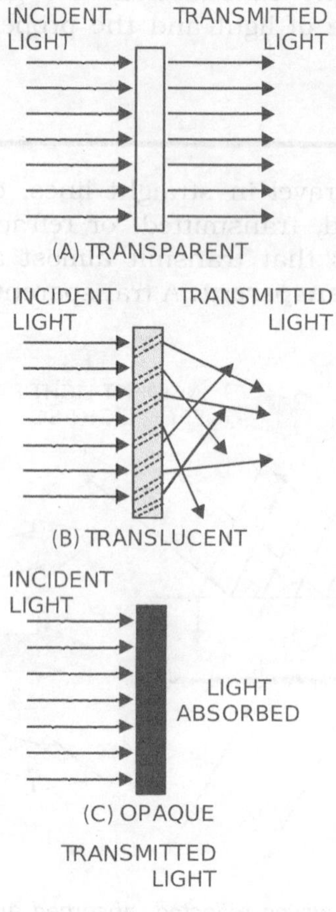

Fig. 2.3. *Substances: A. Transparent; B. Translucent and C. Opaque.*

REFLECTION OF LIGHT

Reflected waves are simply those waves that are neither transmitted nor absorbed, but are reflected from the surface of the medium they encounter. When a wave approaches a reflecting surface, such as a mirror, the wave that strikes the surface is called the **incident** wave, and the one that bounces back is called the **reflected** wave (refer to Fig. 2.4). An imaginary line perpendicular to the point at which the incident wave strikes the reflecting surface is called the **normal,** or the perpendicular. The angle between the incident wave and the normal is called the **angle of incidence.**

The angle between the reflected wave and the normal is called the **angle of reflection.**

If the surface of the medium contacted by the incident wave is smooth and polished, each reflected wave will be reflected back at the same angle as the incident wave. The path of the wave reflected from the surface forms an angle equal to the one formed by its path in reaching the medium.

This conforms to the **law of reflection** which states: The angle of incidence is equal to the angle of reflection.

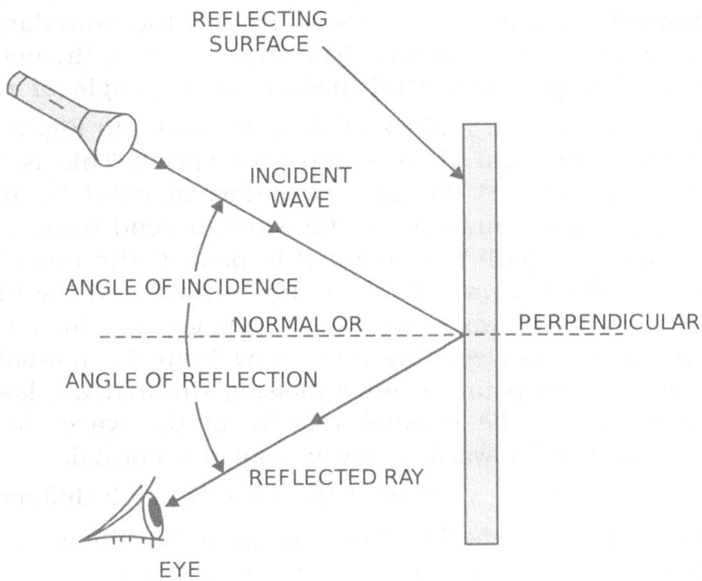

Fig. 2.4. *Reflection of a wave.*

The amount of incident-wave energy that is reflected from a surface depends on the nature of the surface and the angle at which the wave strikes the surface. The amount of wave energy reflected increases as the angle of incidence increases. The reflection of energy is the greatest when the wave is nearly parallel to the reflecting surface. When the incidence wave is perpendicular to the surface, more of the energy is transmitted into the substance and reflection of energy is at its least. At any incident angle, a mirror reflects almost all of the wave energy, while a dull, black surface reflects very little.

Light waves obey the law of reflection. Light travels in a straight line through a substance of uniform density. For example, you can see the straight path of light rays admitted through a narrow slit into a darkened room. The straight path of the beam is made visible by illuminated dust particles suspended in the air. If the light is made to fall onto the surface of a mirror or other reflecting surface, however, the direction of the beam changes sharply.

The light can be reflected in almost any direction, depending on the angle with which the mirror is held.

REFRACTION OF LIGHT

When a light wave passes from one medium into a medium having a different velocity of propagation (the speed waves can travel through a medium), a change in the direction of the wave will occur. This change of direction as the wave enters the second medium is called **refraction.** As in the discussion of reflection, the wave striking the boundary (surface) is called the **incident wave,** and the imaginary line perpendicular to the boundary is called the **normal.** The angle between the incident wave and the normal is called the **angle of incidence.** As the wave passes through the boundary, it is bent either toward or away from the normal. The angle between the normal and the path of the wave through the second medium is the **angle of refraction.**

A light wave passing through a block of glass is shown in Fig. 2.5. The wave moves from point A to point B at a constant speed. This is the incident wave. As the wave penetrates the glass boundary at point B, the velocity of the wave is slowed down. This causes the wave to bend toward the normal. The wave then takes the path from point B to point C through the glass and becomes both the refracted wave from the top surface and the incident wave to the lower surface. As the wave passes from the glass to the air (the second boundary), it is again refracted, this time away from the normal, and takes the path from point C to point D. After passing through the last boundary, the velocity increases to the original velocity of the wave. As illustrated, refracted waves can bend toward or away from the normal.

This bending depends on the velocity of the wave through different mediums.

The broken line between points B and E is the path that the wave would travel if the two mediums (air and glass) had the same density.

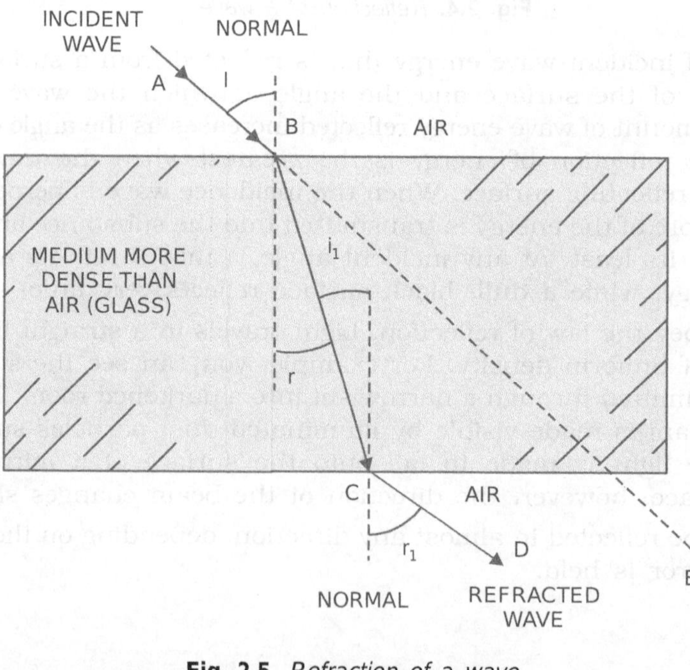

Fig. 2.5. *Refraction of a wave.*

Another interesting condition can be shown using Fig. 2.5. If the wave passes from a less dense to a denser medium, it is bent toward the normal, and the angle of refraction (r) is less than the angle of incidence (i). Likewise, if the wave passes from a denser to a less dense medium, it is bent away from the normal, and the angle of refraction (r_1) is greater than the angle of incidence (i_1).

An example of refraction is the apparent bending of a spoon when it is immersed in a cup of water. The bending seems to take place at the surface of the water or exactly at the point where there is a change of density.

Obviously, the spoon does not bend from the pressure of the water. The light forming the image of the spoon is bent as it passes from the water (a medium of high density) to the air (a medium of comparatively low density).

Without refraction, light waves would pass in straight lines through transparent substances without any change of direction. Figure 2.5 shows that rays striking the glass at any angle other than perpendicular are refracted. However, perpendicular rays, which enter the glass normal to the surface, continue through the glass and into the air in a straight line—no refraction takes place.

DIFFUSION OF LIGHT

When light is reflected from a mirror, the angle of reflection equals the angle of incidence. When light is reflected from a piece of plain white paper; however, the reflected beam is scattered or **diffused,** as shown in Fig. 2.6. Because the surface of the paper is not smooth, the reflected light is broken up into many light beams that are reflected in all directions.

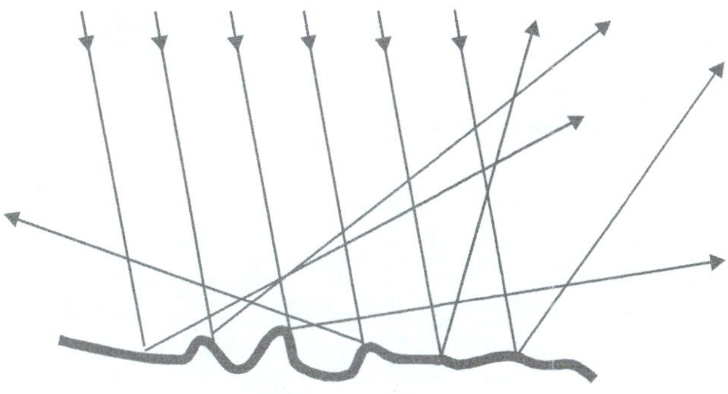

Fig. 2.6. *Diffusion of light.*

ABSORPTION OF LIGHT

You have just seen that a light beam is reflected and diffused when it falls onto a piece of white paper. If the light beam falls onto a piece of black paper, the black paper **absorbs** most of the light rays and very little light is reflected from the paper.

If the surface upon which the light beam falls is perfectly black, there is no reflection; that is, the light is totally absorbed. No matter what kind of surface light falls upon, some of the light is absorbed.

TRANSMISSION OF LIGHT THROUGH OPTICAL FIBERS

The transmission of light along optical fibers depends not only on the nature of light, but also on the structure of the optical fiber. Two methods are used to describe how light is transmitted along the optical fiber. The first method, **ray theory,** uses the concepts of light reflection and refraction. The second method, **mode theory,** treats light as electromagnetic waves. You must first understand the basic optical properties of the materials used to make optical fibers. These properties affect how light is transmitted through the fiber.

BASIC OPTICAL-MATERIAL PROPERTIES

The basic optical property of a material, relevant to optical fibers, is the **index of refraction.** The index of refraction (*n*) measures the speed of light in an optical medium. The index of refraction of a material is the ratio of the speed of light in a vacuum to the speed of light in the material itself. The speed of light (c) in free space (vacuum) is 3×10^8 meters per second (m/s). The speed of light is the frequency (f) of light multiplied by the wavelength of light (λ). When light enters the fiber material (an optically dense medium), the light travels slower at a speed (v). Light will always travel slower in the fiber material than in air. The index of refraction is given by:

$$n = \frac{c}{v}$$

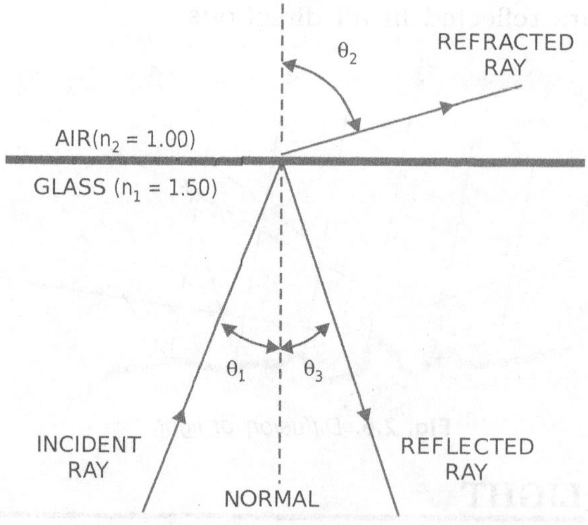

Fig. 2.7 *Light reflection and refraction at a glass-air boundary.*

A light ray is reflected and refracted when it encounters the boundary between two different transparent mediums. For example, Fig. 2.7 shows what happens to the light ray when it encounters the interface between glass and air. The index of refraction for glass (n_1) is 1.50. The index of refraction for air (n_2) is 1.00.

Let's assume the light ray or incident ray is travelling through the glass. When the light ray encounters the glass-air boundary, there are two results. The first result is that part of the ray is reflected back into the glass. The second result is that part of the ray is refracted (bent) as it enters the air. The bending of the light at the glass-air interface is the result of the difference between the index of refractions. Since n_1 is greater than n_2, the angle of refraction (θ_2) will be greater than the angle of incidence (θ_1). Snell's law of refraction is used to describe the relationship between the incident and the refracted rays at the boundary. **Snell's law** is given by:

$$n_1 \times \sin\theta_1 = n_2 \times \sin\theta_2$$

As the angle of incidence θ_1 becomes larger, the angle of refraction (θ_2) approaches 90°. At this point, no refraction is possible. The light ray is totally reflected back into the glass medium. No light escapes into the air. This condition is called **total internal reflection.**

The angle at which total internal reflection occurs is called the **critical angle** of incidence. The critical angle of incidence (θ_c) is shown in Fig. 2.8. At any angle of incidence (θ_2) greater than the critical angle, light is totally reflected back into the glass medium. The critical angle of incidence is determined by using Snell's law. The critical angle is given by:

$$\sin\theta_c = \frac{n_2}{n_1}$$

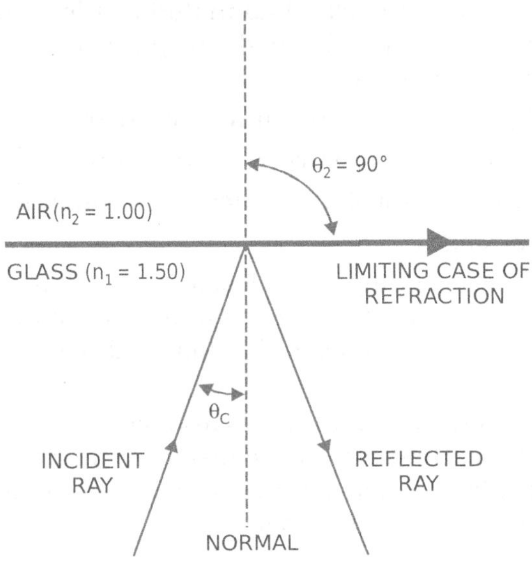

Fig. 2.8. *Critical angle of incidence.*

The condition of total internal reflection is an ideal situation.

However, in reality, there is always some light energy that penetrates the boundary. This situation is explained by the mode theory or the electromagnetic wave theory of light.

BASIC STRUCTURE OF AN OPTICAL FIBER

The basic structure of an optical fiber consists of three parts: the **core,** the **cladding** and the **coating** or **buffer.** The basic structure of an optical fiber is shown in Fig. 2.9. The **core** is a cylindrical rod of dielectric material. Dielectric material conducts no electricity. Light propagates mainly along the core of the fiber. The core is generally made of glass. The **core** is described as having a radius of (*a*) and an index of refraction n_1. The core is surrounded by a layer of material called the **cladding.** Even though light will propagate along the fiber core without the layer of cladding material, the cladding does perform some necessary functions.

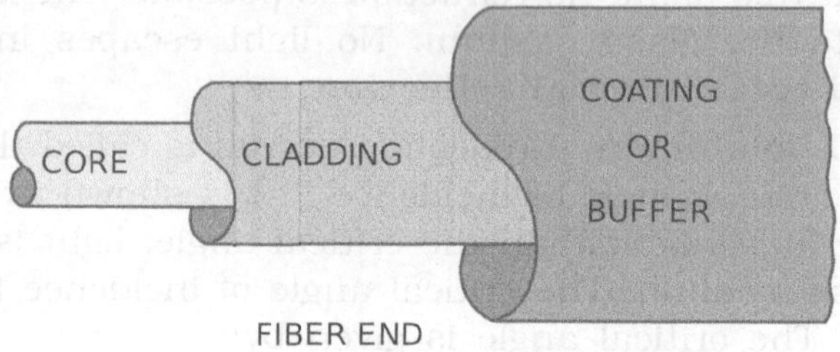

Fig. 2.9. *Basic structure of an optical fiber.*

The **cladding** layer is made of a dielectric material with an index of refraction n_2. The index of refraction of the cladding material is less than that of the core material. The cladding is generally made of glass or plastic. The cladding performs the following functions:

- Reduces loss of light from the core into the surrounding air.
- Reduces scattering loss at the surface of the core.
- Protects the fiber from absorbing surface contaminants.
- Adds mechanical strength.

For extra protection, the cladding is enclosed in an additional layer called the **coating** or **buffer.** The coating or buffer is a layer of material used to protect an optical fiber from physical damage. The material used for a buffer is a type of plastic.

The buffer is elastic in nature and prevents abrasions. The buffer also prevents the optical fiber from scattering losses caused by micro bends. Micro bends occur when an optical fiber is placed on a rough and distorted surface. Micro bends are discussed later in this chapter.

PROPAGATION OF LIGHT ALONG A FIBER

The concept of light propagation, the transmission of light along an optical fiber, can be described by two theories. According to the first theory, light is described as a simple ray. This theory is the ray theory, or geometrical optics, approach. The advantage of the ray approach is that you get a clearer picture of the propagation of light along a fiber. The ray theory is used to approximate the light acceptance and guiding properties of optical fibers. According to the second theory, light is described as an electromagnetic wave. This theory is the mode theory, or wave representation, approach. The mode theory describes the behaviour of light within an optical fiber. The mode theory is useful in describing the optical fiber properties of absorption, attenuation, and dispersion. These fiber properties are discussed later in this chapter.

RAY THEORY

Two types of rays can propagate along an optical fiber. The first type is called **meridional rays**. Meridional rays are rays that pass through the axis of the optical fiber. Meridional rays are used to illustrate the basic transmission properties of optical fibers.

The second type is called **skew rays**. Skew rays are rays that travel through an optical fiber without passing through its axis.

Meridional Rays. Meridional rays can be classified as bound or unbound rays. Bound rays remain in the core and propagate along the axis of the fiber. **Bound rays** propagate through the fiber by total internal reflection. **Unbound rays** are refracted out of the fiber core. Figure 2.10 shows a possible path taken by bound and unbound rays in a step-index fiber. The core of the step-index fiber has an index of refraction n_1. The cladding of a step-index has an index of refraction n_2 that is lower than n_1. Figure 2.10 assumes the core-cladding interface is perfect. However, imperfections at the core-cladding interface will cause part of the bound rays to be refracted out of the core into the cladding. The light rays refracted into the cladding will eventually escape from the fiber. In general, meridional rays follow the laws of reflection and refraction.

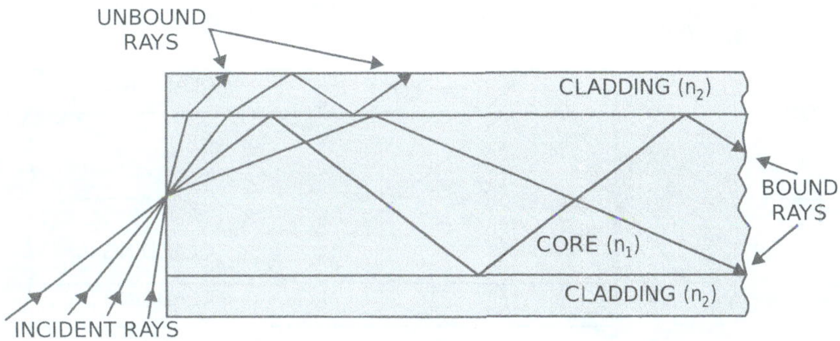

Fig. 2.10. *Bound and unbound rays in a step-index fiber.*

It is known that bound rays propagate in fibers due to total internal reflection, but how do these light rays enter the fiber? Rays that enter the fiber must intersect the core-cladding interface at an angle greater than the critical angle (θ_c). Only those rays that enter the fiber and strike the interface at these angles will propagate along the fiber.

How a light ray is launched into a fiber is shown in Fig. 2.11. The incident ray I_1 enters the fiber at the angle θ_a. I_1 is refracted upon entering the fiber and is transmitted to the core-cladding interface. The ray then strikes the core-cladding interface at the critical angle (θ_c). I_1 is totally reflected back into the core and continues to propagate along the fiber. The incident ray I_2 enters the fiber at an angle greater than θ_a. Again, I_2 is refracted upon entering the fiber and is transmitted to the core-cladding interface. I_2 strikes the core-cladding interface at an angle less than the critical angle (θ_c). I_2 is refracted into the cladding and is eventually lost. The light ray incident on the fiber core must be within the **acceptance cone** defined by the angle θ_a shown in Fig. 2.12.

Angle θ_a is defined as the acceptance angle. The **acceptance angle** (θ_a) is the maximum angle to the axis of the fiber that light entering the fiber is propagated. The value of the angle of acceptance (θ_a) depends on fiber properties and transmission conditions.

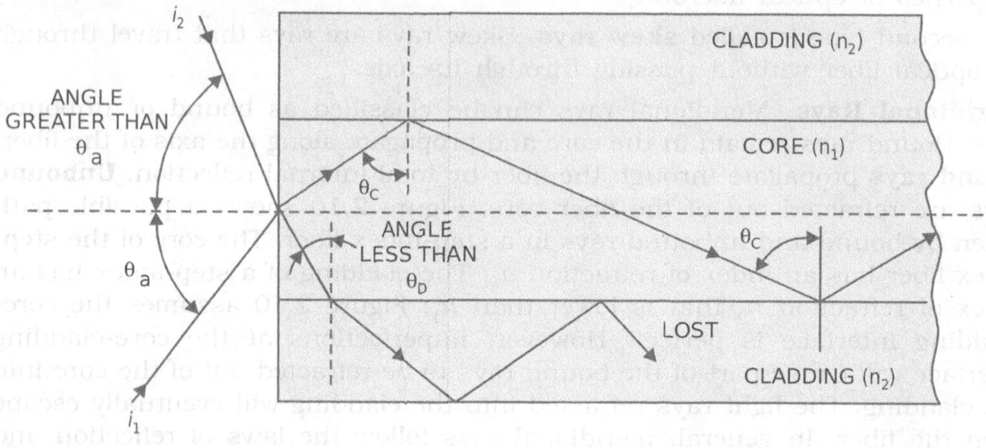

Figure 2.11. *How a light ray enters an optical fiber.*

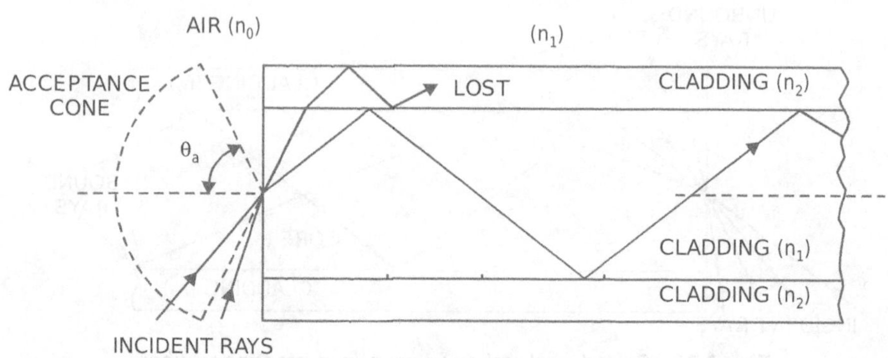

Fig. 2.12. *Fiber acceptance angle.*

The acceptance angle is related to the refractive indices of the core, cladding, and medium surrounding the fiber. This relationship is called the numerical aperture of the fiber. **The numerical aperture (NA) is a measurement of the ability of an optical fiber to capture light. The NA is also used to define the acceptance angle of an optical fiber.**

Figure 2.12 illustrates the relationship between the acceptance angle and the refractive indices. The index of refraction of the fiber core is n_1. The index of refraction of the fiber cladding is n_2. The index of refraction of the surrounding medium is n_0. By using Snell's law and basic trigonometric relationships, the NA of the fiber is given by:

$$NA = n_0 \times \sin \theta_a = (n_1^2 - n_2^{2\,1/2})$$

Since the medium next to the fiber at the launching point is normally air, n_0 is equal to 1.00. The NA is then simply equal to sin θ_a.

The NA is a convenient way to measure the light-gathering ability of an optical fiber. It is used to measure source-to-fiber power-coupling efficiencies. A high NA indicates a high source-to-fiber coupling efficiency.

Source-to-fiber coupling efficiency is described in Chapter 6. Typical values of NA range from 0.20 to 0.29 for glass fibers. Plastic fibers generally have a higher NA. An NA for plastic fibers can be higher than 0.50.

In addition, the NA is commonly used to specify multimode fibers.

However, for small core diameters, such as in single mode fibers, the ray theory breaks down. Ray theory describes only the direction a plane wave takes in a fiber. Ray theory eliminates any properties of the plane wave that interfere with the transmission of light along a fiber. In reality, plane waves interfere with each other. Therefore, only certain types of rays are able to propagate in an optical fiber. Optical fibers can support only a specific number of guided modes. In small core fibers, the number of modes supported is one or only a few modes. Mode theory is used to describe the types of plane waves able to propagate along an optical fiber.

Skew Rays. A possible path of propagation of skew rays is shown in Fig. 2.13. Figure 2.13, view A, provides an angled view and view B provides a front view. Skew rays propagate without passing through the center axis of the fiber.

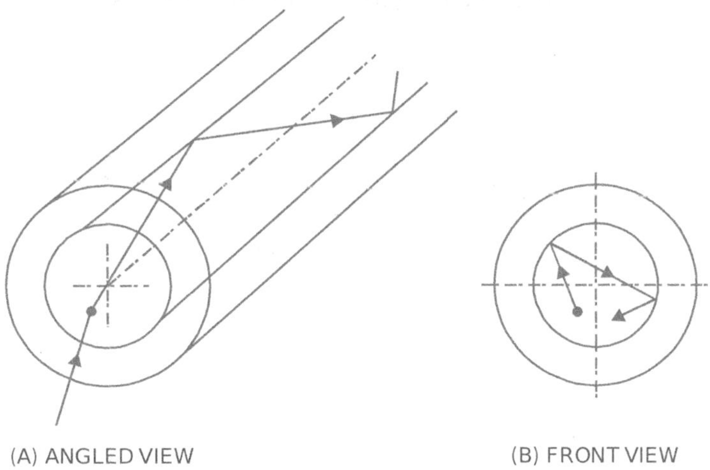

Fig. 2.13. *Skew ray propagation: A. Angled view; B. Front view.*

The acceptance angle for skew rays is larger than the acceptance angle of meridional rays. This condition explains why skew rays outnumber meridional rays. Skew rays are often used in the calculation of light acceptance in an optical fiber. The addition of skew rays increases the amount of light capacity of a fiber. In large NA fibers, the increase may be significant.

The addition of skew rays also increases the amount of loss in a fiber. Skew rays tend to propagate near the edge of the fiber core. A large portion of the number of skew rays that are trapped in the fiber core are considered to be **leaky rays.** Leaky rays are predicted to be totally reflected at the core-cladding boundary. However, these rays are partially refracted because of the curved nature of the fiber boundary. Mode theory is also used to describe this type of leaky ray loss.

MODE THEORY

The mode theory, along with the ray theory, is used to describe the propagation of light along an optical fiber. The mode theory is used to describe the properties of light that ray theory is unable to explain. The mode theory uses electromagnetic wave behavior to describe the propagation of light along a fiber. A set of guided electromagnetic waves is called the **modes** of the fiber.

Plane Waves. The mode theory suggests that a light wave can be represented as a plane wave. A **plane wave** is described by its direction, amplitude, and wavelength of propagation. A plane wave is a wave whose surfaces of constant phase are infinite parallel planes normal to the direction of propagation.

The planes having the same phase are called the wavefronts. The wavelength (λ) of the plane wave is given by: wavelength (λ) = $\frac{c}{fn}$, where c is the speed of light in a vacuum, f is the frequency of the light and n is the index of refraction of the plane-wave medium.

Figure 2.14 shows the direction and wavefronts of plane-wave propagation. Plane waves, or wavefronts, propagate along the fiber similar to light rays.

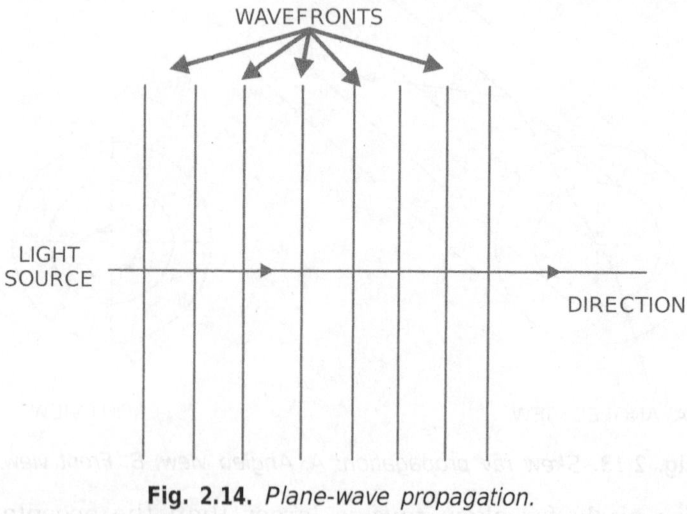

Fig. 2.14. *Plane-wave propagation.*

However, not all wavefronts incident on the fiber at angles less than or equal to the critical angle of light acceptance propagate along the fiber. Wavefronts may undergo a change in phase that prevents the successful transfer of light along the fiber.

Wavefronts are required to remain in phase for light to be transmitted along the fiber. Consider the wavefront incident on the core of an optical fiber as shown in Fig. 2.15. Only those wavefronts incident on the fiber at angles less than or equal to the critical angle may propagate along the fiber. The wavefront undergoes a gradual phase change as it travels down the fiber. Phase changes also occur when the wavefront is reflected. The wavefront must remain in phase after the wavefront transverses the fiber twice and is reflected twice. The distance transversed is shown between point A and point B on Fig. 2.15. The reflected waves at point A and point B are in phase if the total amount of phase collected is an integer multiple of 2π radian. If propagating wavefronts are not in phase, they eventually disappear. Wavefronts disappear because of **destructive interference.** The wavefronts that are in phase interfere with the wavefronts that are out of phase. This interference is the reason why only a finite number of modes can propagate along the fiber.

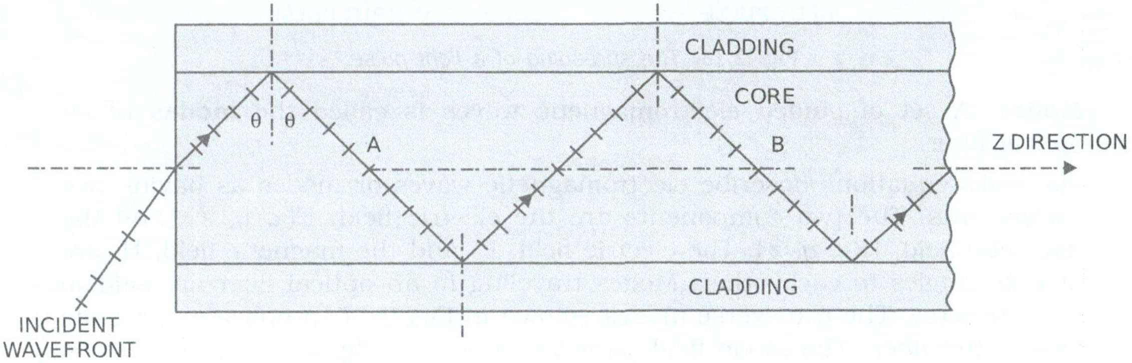

Fig. 2.15. *Wavefront propagation along an optical fiber.*

The plane waves repeat as they travel along the fiber axis. The direction the plane waves travel is assumed to be the z direction as shown in Fig. 2.15. The plane waves repeat at a distance equal to $\lambda/\sin\theta$. Plane waves also repeat at a periodic frequency $\beta_2 = 2\pi\sin\theta/\lambda$. The quantity β is defined as the **propagation constant** along the fiber axis. As the wavelength (λ) changes, the value of the propagation constant must also change.

For a given mode, a change in wavelength can prevent the mode from propagating along the fiber. The mode is no longer bound to the fiber. The mode is said to be cut off. Modes that are bound at one wavelength may not exist at longer wavelengths. The wavelength at which a mode ceases to be bound is called the **cutoff wavelength** for that mode. However, an optical fiber is always able to propagate at least one mode. This mode is referred to as the fundamental mode of the fiber. The fundamental mode can never be cutoff.

The wavelength that prevents the next higher mode from propagating is called the cutoff wavelength of the fiber. An optical fiber that operates above the cutoff wavelength (at a longer wavelength) is called a single mode fiber. An optical fiber that operates below the cutoff wavelength is called a multimode fiber. Single mode and multimode optical fibers are discussed later in this chapter.

In a fiber, the propagation constant of a plane wave is a function of the wave's wavelength and mode. The change in the propagation constant for different waves is called **dispersion.** The change in the propagation constant for different wavelengths is called **chromatic dispersion.** The change in propagation constant for different modes is called **modal dispersion.**

These dispersions cause the light pulse to spread as it goes down the fiber (Fig. 2.16). Some dispersion occurs in all types of fibers. Dispersion is discussed later in this chapter.

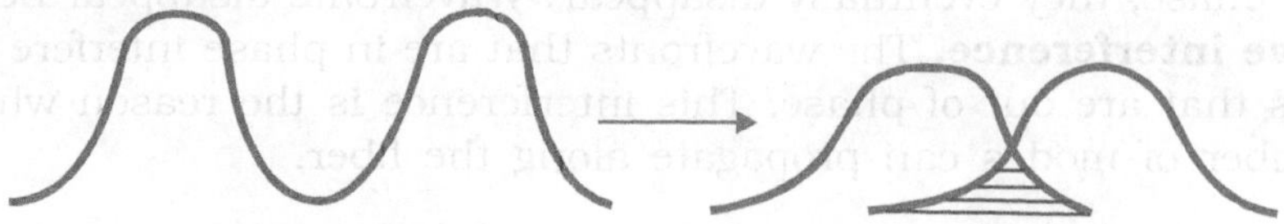

Fig. 2.16. *The spreading of a light pulse.*

Modes. A set of guided electromagnetic waves is called the **modes** of an optical fiber.

Maxwell's equations describe electromagnetic waves or modes as having two components. The two components are the electric field, E(x, y, z), and the magnetic field, H(x, y, z). The electric field, E, and the magnetic field, H, are at right angles to each other. Modes traveling in an optical fiber are said to be transverse. The transverse modes, shown in Fig. 2.17, propagate along the axis of the fiber. The mode field patterns shown in Fig. 2.17 are said to be transverse electric (TE). In TE modes, the electric field is perpendicular to the direction of propagation.

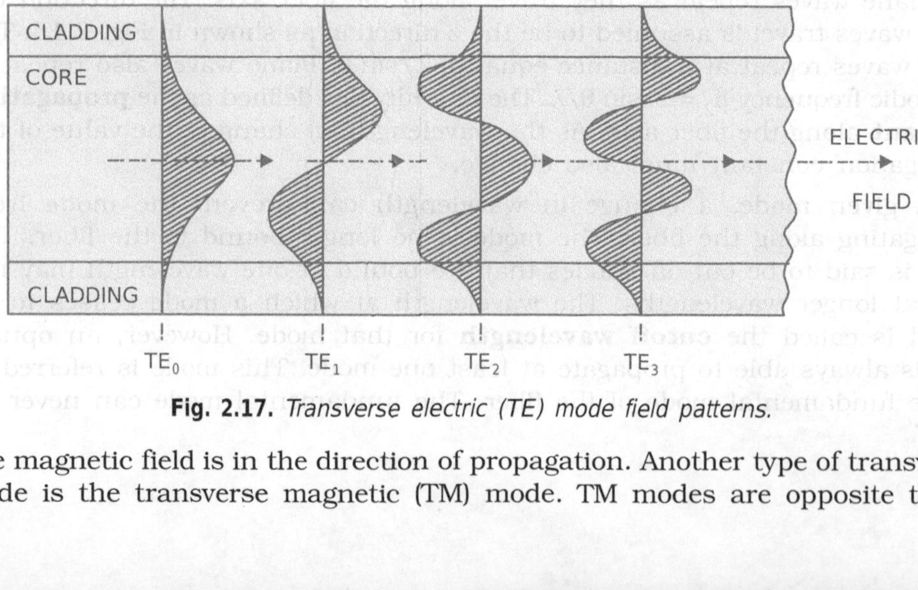

Fig. 2.17. *Transverse electric (TE) mode field patterns.*

The magnetic field is in the direction of propagation. Another type of transverse mode is the transverse magnetic (TM) mode. TM modes are opposite to TE modes. In TM modes, the magnetic field is perpendicular to the direction of propagation. The electric field is in the direction of propagation. Figure 2.17 shows only TE modes.

The TE mode field patterns shown in Fig. 2.17 indicate the order of each mode. The order of each mode is indicated by the number of field maxima within the core of the fiber. For example, TE_0 has one field maxima. The electric field is a maximum at the center of the waveguide and decays toward the core-cladding boundary. TE_0 is considered the fundamental mode or the lowest order standing wave. As the number of field maxima increases, the order of the mode is higher. Generally, modes with more than a few (5–10) field maxima are referred to as high-order modes.

The order of the mode is also determined by the angle the wavefront makes with the axis of the fiber. Figure 2.18 illustrates light rays as they travel down the fiber. These light rays indicate the direction of the wavefronts. High-order modes cross the axis of the fiber at steeper angles. Low-order and high-order modes are shown in Fig. 2.18.

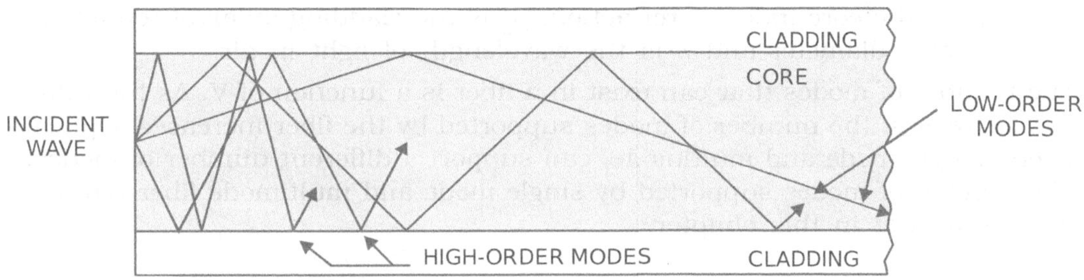

Fig. 2.18. *Low-order and high-order modes.*

Before we progress, let us refer back to Fig. 2.17.

Notice that the modes are not confined to the core of the fiber. The modes extend partially into the cladding material. Low-order modes penetrate the cladding only slightly. In low-order modes, the electric and magnetic fields are concentrated near the center of the fiber. However, high-order modes penetrate further into the cladding material. In high-order modes, the electrical and magnetic fields are distributed more toward the outer edges of the fiber.

This penetration of low-order and high-order modes into the cladding region indicates that some portion is refracted out of the core. The refracted modes may become trapped in the cladding due to the dimension of the cladding region. The modes trapped in the cladding region are called **cladding modes.** As the core and the cladding modes travel along the fiber, mode coupling occurs. **Mode coupling** is the exchange of power between two modes. Mode coupling to the cladding results in the loss of power from the core modes.

In addition to bound and refracted modes, there are leaky modes.

Leaky modes are similar to leaky rays. **Leaky modes** lose power as they propagate along the fiber. For a mode to remain within the core, the mode must meet certain boundary conditions. A mode remains bound if the propagation constant β meets the following boundary condition:

$$\frac{2\pi n_2}{\lambda} < \beta < \frac{2\pi n_1}{\lambda}$$

where n_1 and n_2 are the index of refraction for the core and the cladding, respectively. When the propagation constant becomes smaller than $2\pi n_2/\lambda$, power leaks out of the core and into the cladding. Generally, modes leaked into the cladding are lost in a few centimeters. However, leaky modes can carry a large amount of power in short fibers.

Normalized Frequency. Electromagnetic waves bound to an optical fiber are described by the fiber's normalized frequency.

The **normalized frequency** determines how many modes a fiber can support. Normalized frequency is a dimensionless quantity.

Normalized frequency is also related to the fiber's cutoff wavelength. Normalized frequency (V) is defined as:

$$V = \frac{2\pi a}{\lambda}\left(n_1^2 - n_2^2\right)^{1/2}$$

where n_1 is the core index of refraction, n_2 is the cladding index of refraction, α is the core diameter and λ is the wavelength of light in air.

The number of modes that can exist in a fiber is a function of V. As the value of V increases, the number of modes supported by the fiber increases. Optical fibers, single mode and multimode, can support a different number of modes. The number of modes supported by single mode and multimode fiber types is discussed later in this chapter.

OPTICAL FIBER TYPES

Optical fibers are characterized by their structure and by their properties of transmission. Basically, optical fibers are classified into two types. The first type is *single mode fibers*. The second type is *multimode fibers*. As each name implies, optical fibers are classified by the number of modes that propagate along the fiber. As previously explained, the structure of the fiber can permit or restrict modes from propagating in a fiber. The basic structural difference is the core size. Single mode fibers are manufactured with the same materials as multimode fibers. Single mode fibers are also manufactured by following the same fabrication process as multimode fibers.

Single Mode Fibers

The core size of single mode fibers is small. The core size (diameter) is typically around 8 to 10 micrometers (μm). A fiber core of this size allows only the fundamental or lowest order mode to propagate around a 1300 nanometer (nm) wavelength. Single mode fibers propagate only one mode, because the core size approaches the operational wavelength (λ). The value of the normalized frequency parameter (V) relates core size with mode propagation.

In single mode fibers, V is less than or equal to 2.405. When $V \leq 2.405$, single mode fibers propagate the fundamental mode down the fiber core, while high-order modes are lost in the cladding. For low V values (≤ 1.0), most of the power is propagated in the cladding material. Power transmitted by the cladding is easily lost at fiber bends. The value of V should remain near the 2.405 level.

Single mode fibers have a lower signal loss and a higher information capacity (bandwidth) than multimode fibers. Single mode fibers are capable of transferring higher amounts of data due to low fiber dispersion. Basically, dispersion is the spreading of light as light propagates along a fiber. Dispersion mechanisms in single mode fibers are discussed in more detail later in this chapter. Signal loss depends on the operational wavelength (λ). In single mode fibers, the wavelength can increase or decrease the losses caused by fiber bending. Single mode fibers operating at wavelengths larger than the cut-off wavelength lose more power at fiber bends. They lose power because light radiates into the cladding, which is lost at fiber bends. In general, single mode fibers are considered to be low-loss fibers, which increase system bandwidth and length.

Multimode Fibers

As their name implies, multimode fibers propagate more than one mode. Multimode fibers can propagate over 100 modes. The number of modes propagated depends on the core size and numerical aperture (NA). As the core size and NA increase, the number of modes increases. Typical values of fiber core size and NA are 50 to 100 μm and 0.20 to 0.29, respectively.

A large core size and a higher NA have several advantages. Light is launched into a multimode fiber with more ease. The higher NA and the larger core size make it easier to make fiber connections. During fiber splicing, core-to-core alignment becomes less critical. Another advantage is that multimode fibers permit the use of light-emitting diodes (LEDs). Single mode fibers typically must use laser diodes. LEDs are cheaper, less complex, and last longer. LEDs are preferred for most applications.

Multimode fibers also have some disadvantages. As the number of modes increases, the effect of modal dispersion increases. Modal dispersion (intermodal dispersion) means that modes arrive at the fiber end at slightly different times. This time difference causes the light pulse to spread. Modal dispersion affects system bandwidth. Fiber manufacturers adjust the core diameter, NA, and index profile properties of multimode fibers to maximize system bandwidth.

PROPERTIES OF OPTICAL FIBER TRANSMISSION

The principles behind the transfer of light along an optical fiber were discussed earlier in this chapter. You learned that propagation of light depended on the nature of light and the structure of the optical fiber. However, our discussion did not describe how optical fibers affect system performance.

In this case, system performance deals with signal loss and bandwidth.

Signal loss and system bandwidth describe the amount of data transmitted over a specified length of fiber. Many optical fiber properties increase signal loss and reduce system bandwidth. The most important properties that affect system performance are fiber attenuation and dispersion.

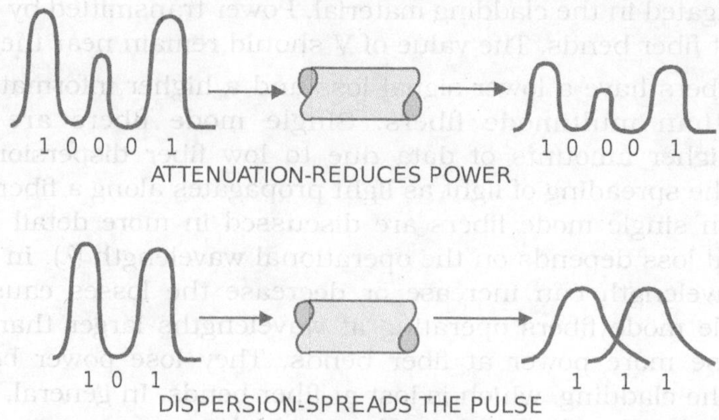

Fig. 2.19. *Fiber transmission properties.*

Attenuation reduces the amount of optical power transmitted by the fiber. Attenuation controls the distance an optical signal (pulse) can travel as shown in Fig. 2.19. Once the power of an optical pulse is reduced to a point where the receiver is unable to detect the pulse, an error occurs. Attenuation is mainly a result of **light absorption, scattering,** and **bending losses.** Dispersion spreads the optical pulse as it travels along the fiber. This spreading of the signal pulse reduces the system bandwidth or the information-carrying capacity of the fiber. Dispersion limits how fast information is transferred as shown in Fig. 2.19. An error occurs when the receiver is unable to distinguish between input pulses caused by the spreading of each pulse. The effects of attenuation and dispersion increase as the pulse travels the length of the fiber as shown in Fig. 2.20.

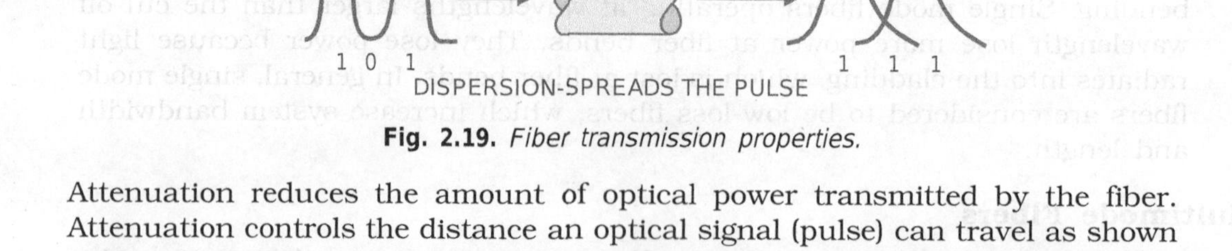

Fig. 2.20. *Pulse spreading and power loss along an optical fiber.*

In addition to fiber attenuation and dispersion, other optical fiber properties affect system performance. Fiber properties, such as modal noise, pulse broadening, and polarization, can reduce system performance.

Modal noise, pulse broadening, and polarization are too complex to discuss as introductory level material. However, you should be aware that attenuation and dispersion are not the only fiber properties that affect performance.

Attenuation

Attenuation in an optical fiber is caused by absorption, scattering, and bending losses. Attenuation is the loss of optical power as light travels along the fiber. Signal attenuation is defined as the ratio of optical input power (P_i) to the optical output power (P_o). Optical input power is the power injected into the fiber from an optical source. Optical output power is the power received at the fiber end or optical detector. The following equation defines signal attenuation as a unit of length:

$$\text{Attenuation} = \left(\frac{10}{L}\right) \log_{10}\left(\frac{P_i}{P_o}\right)$$

Signal attenuation is a log relationship. Length (L) is expressed in kilometers. Therefore, the unit of attenuation is decibels/kilometer (dB/km). As previously stated, attenuation is caused by absorption, scattering, and bending losses. Each mechanism of loss is influenced by fiber-material properties and fiber structure. However, loss is also present at fiber connections. Fiber connector, splice, and coupler losses are discussed in Chapter 4. The present discussion remains relative to optical fiber attenuation properties.

Absorption. Absorption is a major cause of signal loss in an optical fiber. Absorption is defined as the portion of attenuation resulting from the conversion of optical power into another energy form, such as heat. Absorption in optical fibers is explained by three factors:

- Imperfections in the atomic structure of the fiber material.
- The intrinsic or basic fiber-material properties.
- The extrinsic (presence of impurities) fiber-material properties.

Imperfections in the atomic structure induce absorption by the presence of missing molecules or oxygen defects. Absorption is also induced by the diffusion of hydrogen molecules into the glass fiber. Since intrinsic and extrinsic material properties are the main cause of absorption, they are discussed further.

Intrinsic Absorption. Intrinsic absorption is caused by basic fiber-material properties. If an optical fiber were absolutely pure, with no imperfections or impurities, then all absorption would be intrinsic. Intrinsic absorption sets the minimal level of absorption.

In fiber optics, silica (pure glass) fibers are used predominately. Silica fibers are used because of their low intrinsic material absorption at the wavelengths of operation.

In silica glass, the wavelengths of operation range from 700 nanometers (nm) to 1600 nm. Figure 2.21 shows the level of attenuation at the wavelengths of operation. This wavelength of operation is between two intrinsic absorption regions. The first region is the **ultraviolet** region (below 400 nm wavelength). The second region is the **infrared** region (above 2000 nm wavelength).

Intrinsic absorption in the ultraviolet region is caused by electronic absorption bands. Basically, absorption occurs when a light particle (photon) interacts with an electron and excites it to a higher energy level. The tail of the ultraviolet absorption band is shown in Fig. 2.21.

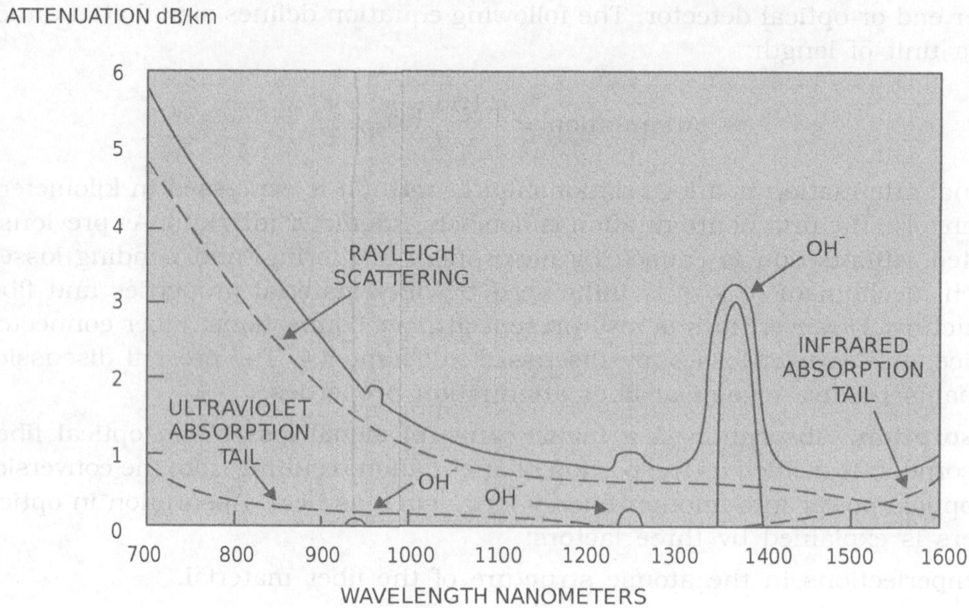

Fig. 2.21. *Fiber losses.*

The main cause of **intrinsic absorption** in the infrared region is the characteristic vibration frequency of atomic bonds. In silica glass, absorption is caused by the vibration of silicon-oxygen (Si-O) bonds. The interaction between the vibrating bond and the electromagnetic field of the optical signal causes intrinsic absorption. Light energy is transferred from the electromagnetic field to the bond. The tail of the infrared absorption band is shown in Fig. 2.21.

Extrinsic Absorption. Extrinsic absorption is caused by impurities introduced into the fiber material. Trace metal impurities, such as iron, nickel, and chromium, are introduced into the fiber during fabrication. Extrinsic absorption is caused by the electronic transition of these metal ions from one energy level to another.

Extrinsic absorption also occurs when hydroxyl ions (OH^-) are introduced into the fiber. Water in silica glass forms a silicon-hydroxyl (Si-OH) bond. This bond has a fundamental absorption at 2700 nm. However, the harmonics or overtones of the fundamental absorption occur in the region of operation. These harmonics increase extrinsic absorption at 1383 nm, 1250 nm, and 950 nm. Figure 2.21 shows the presence of the three OH^- harmonics. The level of the OH^- harmonic absorption is also indicated.

These absorption peaks define three regions or windows of preferred operation. The first window is centered at 850 nm. The second window is centered at 1300 nm. The third window is centered at 1550 nm. Fiber optic systems operate at wavelengths defined by one of these windows.

The amount of water impurities present in a fiber should be less than a few parts per billion. Fiber attenuation caused by extrinsic absorption is affected by the level of impurities (OH^-) present in the fiber. If the amount of impurities in a fiber is reduced, then fiber attenuation is reduced.

Scattering. Basically, scattering losses are caused by the interaction of light with density fluctuations within a fiber. Density changes are produced when optical fibers are manufactured.

During manufacturing, regions of higher and lower molecular density areas, relative to the average density of the fiber, are created. Light travelling through the fiber interacts with the density areas as shown in Fig. 2.22. Light is then partially scattered in all directions.

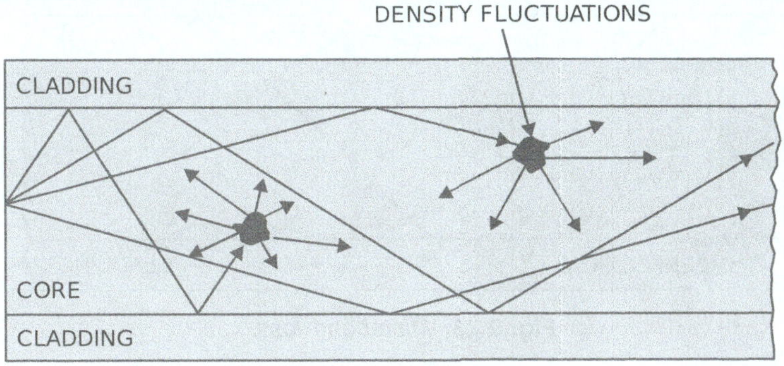

Fig. 2.22. *Light scattering.*

In commercial fibers operating between 700 nm and 1600 nm wavelength, the main source of loss is called **Rayleigh scattering.** Rayleigh scattering is the main loss mechanism between the ultraviolet and infrared regions as shown in Fig. 2.21. Rayleigh scattering occurs when the size of the density fluctuation (fiber defect) is less than one-tenth of the operating wavelength of light. Loss caused by Rayleigh scattering is proportional to the fourth power of the wavelength ($1/\lambda^4$). As the wavelength increases, the loss caused by Rayleigh scattering decreases.

If the size of the defect is greater than one-tenth of the wavelength of light, the scattering mechanism is called **Mie scattering.** Mie scattering, caused by these large defects in the fiber core, scatters light out of the fiber core. However, in commercial fibers, the effects of Mie scattering are insignificant. Optical fibers are manufactured with very few large defects.

Bending Loss. Bending the fiber also causes attenuation. Bending loss is classified according to the bend radius of curvature: microbend loss or macrobend loss.

Microbends are small microscopic bends of the fiber axis that occur mainly when a fiber is cabled. **Macrobends** are bends having a large radius of curvature relative to the fiber diameter. Microbend and macrobend losses are very important loss mechanisms. Fiber loss caused by microbending can still occur even if the fiber is cabled correctly. During installation, if fibers are bent too sharply, macrobend losses will occur.

Microbend losses are caused by small discontinuities or imperfections in the fiber. Uneven coating applications and improper cabling procedures increase microbend loss. External forces are also a source of microbends. An external force deforms the cabled jacket surrounding the fiber but causes only a small bend in the fiber. Microbends change the path that propagating modes take, as shown in Fig. 2.23. **Microbend loss** increases attenuation because low-order modes become coupled with high-order modes that are naturally lossy.

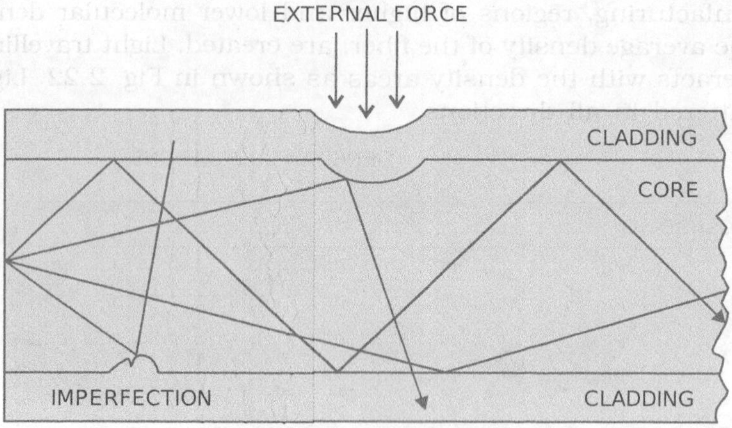

Fig. 2.23. *Microbend loss.*

Macrobend losses are observed when a fiber bend's radius of curvature is large compared to the fiber diameter.

These bends become a great source of loss when the radius of curvature is less than several centimeters. Light propagating at the inner side of the bend travels a shorter distance than that on the outer side. To maintain the phase of the light wave, the mode phase velocity must increase. When the fiber bend is less than some critical radius, the mode phase velocity must increase to a speed greater than the speed of light. However, it is impossible to exceed the speed of light. This condition causes some of the light within the fiber to be converted to high-order modes. These high-order modes are then lost or radiated out of the fiber.

Fiber sensitivity to bending losses can be reduced. If the refractive index of the core is increased, then fiber sensitivity decreases. Sensitivity also decreases as the diameter of the overall fiber increases. However, increases in the fiber core diameter increase fiber sensitivity. Fibers with larger core size propagate more modes. These additional modes tend to be more lossy.

DISPERSION

There are two different types of dispersion in optical fibers.

The types are intramodal and intermodal dispersion. Intramodal, or chromatic, dispersion occurs in all types of fibers. Intermodal or modal, dispersion occurs only in multimode fibers. Each type of dispersion mechanism leads to pulse spreading. As a pulse spreads, energy is overlapped. This condition is shown in Fig. 2.24. The spreading of the optical pulse as it travels along the fiber limits the information capacity of the fiber.

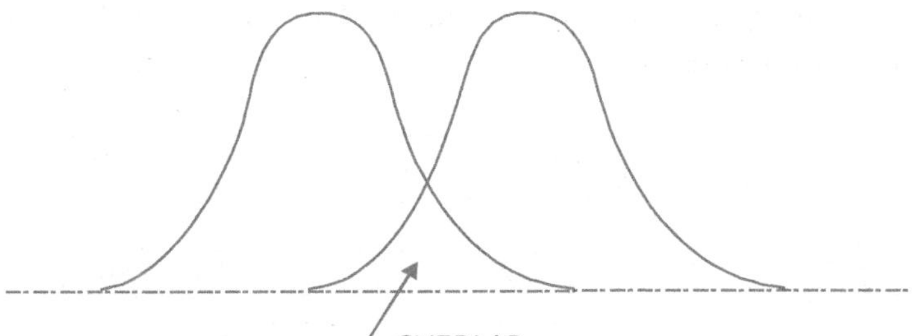

Fig. 2.24. *Pulse overlap.*

Intramodal Dispersion

Intramodal, or chromatic, dispersion depends primarily on fiber materials. There are two types of intramodal dispersion. The first type is material dispersion. The second type is waveguide dispersion.

Intramodal dispersion occurs because different colors of light travel through different materials and different waveguide structures at different speeds.

Material dispersion occurs because the spreading of a light pulse is dependent on the wavelengths' interaction with the refractive index of the fiber core. Different wavelengths travel at different speeds in the fiber material. Different wavelengths of a light pulse that enter a fiber at one time exit the fiber at different times. Material dispersion is a function of the source spectral width. The spectral width specifies the range of wavelengths that can propagate in the fiber. Material dispersion is less at longer wavelengths.

Waveguide dispersion occurs because the mode propagation constant (β) is a function of the size of the fiber's core relative to the wavelength of operation. Waveguide dispersion also occurs because light propagates differently in the core than in the cladding.

In multimode fibers, waveguide dispersion and material dispersion are basically separate properties. Multimode waveguide dispersion is generally small compared to material dispersion. Waveguide dispersion is usually neglected.

However, in single mode fibers, material and waveguide dispersion are interrelated.

The total dispersion present in single mode fibers may be minimized by trading material and waveguide properties depending on the wavelength of operation.

Intermodal Dispersion

Intermodal or modal dispersion causes the input light pulse to spread. The input light pulse is made up of a group of modes. As the modes propagate along the fiber, light energy distributed among the modes is delayed by different amounts. The pulse spreads because each mode propagates along the fiber at different speeds. Since modes travel in different directions, some modes travel longer distances. **Modal dispersion** occurs because each mode travels a different distance over the same time span, as shown in Fig. 2.25. The modes of a light pulse that enter the fiber at one time exit the fiber a different times. This condition causes the light pulse to spread. As the length of the fiber increases, modal dispersion increases.

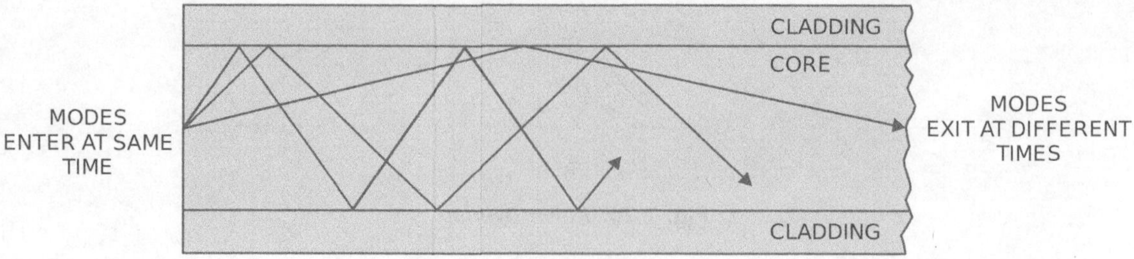

Fig. 2.25. *Distance travelled by each mode over the same time span.*

Modal dispersion is the dominant source of dispersion in multimode fibers. Modal dispersion does not exist in single mode fibers. Single mode fibers propagate only the fundamental mode. Therefore, single mode fibers exhibit the lowest amount of total dispersion. Single mode fibers also exhibit the highest possible bandwidth.

SUMMARY

Now that you have completed this chapter, let's review some of the new terms, concepts, and ideas that you have learned. You should have a thorough understanding of these principles before moving on to Chapter 3.

A LIGHT WAVE is a form of energy that is moved by wave motion.

WAVE MOTION is defined as a recurring disturbance advancing through space with or without the use of a physical medium.

SCIENTIFIC EXPERIMENTS seem to show that light is composed of tiny particles, while other experiments indicate that light is made up of waves. Today, physicists have come to accept a theory concerning light that is a combination of particle (ray) theory and wave (mode) theory.

TRANSVERSE WAVE MOTION describes the up and down wave motion that is at right angle (transverse) to the outward motion of the waves.

LIGHT RAYS, when they encounter any substance, are either transmitted, refracted, reflected, or absorbed.

REFLECTION occurs when a wave strikes an object and bounces back (toward the source). The wave that moves from the source to the object is called the **incident wave,** and the wave that moves away from the object is called the **reflected wave.**

The **LAW OF REFLECTION** states that the angle of incidence is equal to the angle of reflection.

REFRACTION occurs when a wave travelling through two different mediums passes through the **boundary** of the mediums and bends toward or away from the **normal.**

The RAY THEORY and the MODE THEORY describe how light energy is transmitted along an optical fiber.

The INDEX OF REFRACTION is the basic optical material property that measures the speed of light in an optical medium.

SNELL'S LAW OF REFRACTION describes the relationship between the incident and the refracted rays when light rays encounter the boundary between two different transparent materials.

TOTAL INTERNAL REFLECTION occurs when light rays are totally reflected at the boundary between two different transparent materials. The angle at which total internal reflection occurs is called the **critical angle of incidence.**

The CORE, CLADDING and COATING or BUFFER are the three basic parts of an optical fiber.

The RAY THEORY describes how light rays propagate along an optical fiber. MERIDIONAL RAYS pass through the axis of the optical fiber. SKEW RAYS propagate through an optical fiber without passing through its axis.

BOUND RAYS propagate through an optical fiber core by total internal reflection. UNBOUND RAYS refract out of the fiber core into the cladding and are eventually lost.

The ACCEPTANCE ANGLE is the maximum angle to the axis of the fiber that light entering the fiber is bound or propagated.

The light ray incident on the fiber core must be within the acceptance cone defined by the acceptance angle to be propagated along an optical fiber.

NUMERICAL APERTURE (NA) is a measurement of the ability of an optical fiber to capture light.

The MODE THEORY uses electromagnetic wave behavior to describe the propagation of light along an optical fiber. A set of guided electromagnetic waves are called the **modes** of the fiber.

MODES travelling in an optical fiber are said to be transverse. Modes are described by their electric, $E(x, y, z)$, and magnetic, $H(x, y, z)$, fields. The electric field and magnetic field are at right angles to each other.

NORMALIZED FREQUENCY determines how many modes a fiber can support. The number of modes is represented by the normalized frequency constant.

SINGLE MODE and MULTIMODE

FIBERS are classified by the number of modes that propagate along the optical fiber. Single mode fibers propagate only one mode because the core size approaches the operational wavelength. Multimode fibers can propagate over 100 modes depending on the core size and numerical aperture.

ATTENUATION is the loss of optical power as light travels along an optical fiber. Attenuation in an optical fiber is caused by absorption, scattering and bending losses.

DISPERSION spreads the optical pulse as it travels along the fiber. Dispersion limits how fast information is transferred.

ABSORPTION is the conversion of optical power into another energy form, such as heat. INTRINSIC ABSORPTION is caused by basic fiber-material properties. EXTRINSIC ABSORPTION is caused by impurities introduced into the fiber material.

SILICA FIBERS are predominately used in fiber optic communications. They have low intrinsic material absorption at the wavelengths of operation.

The WAVELENGTH OF OPERATION in fiber optics is between 700 nm and 1600 nm. The wavelength of operation is between the ultraviolet (below 400 nm) and infrared (above 2000 nm) intrinsic absorption regions.

EXTRINSIC ABSORPTION occurs when impurities, such as hydroxyl ions (OH^-), are introduced into the fiber. OH^- absorption peaks define three regions or windows of preferred operation. The first window is centered at 850 nm. The second window is centered at 1300 nm. The third window is centered at 1550 nm.

SCATTERING losses are caused by the interaction of light with density fluctuations within a fiber. **Rayleigh scattering** is the main source of loss in commercial fibers operating between 700 nm and 1600 nm.

MICROBENDS are small microscopic bends of the fiber axis that occur mainly when a fiber is cabled. MACROBENDS are bends having a large radius of curvature relative to the fiber diameter.

INTRAMODAL, or CHROMATIC, DISPERSION **occurs because light travels through different materials and different waveguide structures at different speeds.**

MATERIAL DISPERSION is dependent on the light wavelengths interaction with the refractive index of the core. WAVEGUIDE DISPERSION is a function of the size of the fiber's core relative to the wavelength of operation.

INTERMODAL or MODAL, DISPERSION **occurs because each mode travels a different distance over the same time span.**

REVIEW QUESTIONS AND ANSWERS

Q. 1. *Quantum physics successfully explained the photoelectric effect in terms of fundamental particles of energy called quanta. What are the fundamental particles of energy (quanta) known as when referring to light energy?*

Ans. Photons.

Q. 2. *What type of wave motion is represented by the motion of water?*

Ans. Transverse-wave motion.

Q. 3. *When light waves encounter any substance, what four things can happen?*

Ans. Light waves are either transmitted, refracted, reflected or absorbed.

Q. 4. *A substance that transmits almost all of the light waves falling upon it is known as what type of substance?*

Ans. Transparent.

Q. 5. *A substance that is unable to transmit any light waves is known as what type of substance?*

Ans. Opaque.

Q. 6. *What is the law of reflection?*

Ans. The law of reflection states that the angle of incidence is equal to the angle of reflection.

Q. 7. *When a wave is reflected from a surface, energy is reflected. When is the reflection of energy the greatest?*

Ans. When the wave is nearly parallel to the reflecting surface.

Q. 8. *When is the reflection energy the least?*

Ans. When the wave is perpendicular to the reflecting surface.

Q. 9. *Light waves obey what law?*

Ans. The law of reflection.

Q. 10. *A refracted wave occurs when a wave passes from one medium into another medium. What determines the angle of refraction?*

Ans. Depends on the bending caused by the velocity difference of the wave travelling through different mediums.

Q. 11. *A light wave enters a sheet of glass at a perfect right angle to the surface. Is the majority of the wave reflected, refracted, transmitted or absorbed?*

Ans. Transmitted.

Q. 12. *When light strikes a piece of white paper, the light is reflected in all directions. What do we call this scattering of light?*

Ans. Diffusion.

Q. 13. *Two methods describe how light propagates along an optical fiber. These methods define two theories of light propagation. What do we call these two theories?*

Ans. The ray theory and the mode theory.

Q. 14. *What is the basic optical-material property relevant to optical fiber light transmission ?*

Ans. The index of refraction.

Q. 15. *The index of refraction measures the speed of light in an optical fiber. Will light travel faster in an optically dense material or in one that is less dense?*

Ans. Light will travel faster in an optical material that is less dense.

Q. 16. *Assume light is travelling through glass, what happens when this light strikes the glass-air boundary?*

Ans. Part of the light ray is reflected back into the glass and part of the light ray is refracted (bent) as it enters the air.

Q. 17. *What condition causes a light ray to be totally reflected back into its medium of propagation?*

Ans. Total internal reflection occurs when the angle of refraction approaches 90°. This condition occurs when the angle of incidence increases to the point where no refraction is possible.

Q. 18. *What name is given to the angle where total internal reflection occurs?*

Ans. Critical angle of incidence.

Q. 19. *List the three parts of an optical fiber.*

Ans. Core, cladding, and coating or buffer.

Q. 20. *Which fiber material, core or cladding, has a higher index of refraction?*

Ans. Core.

Q. 21. *Light transmission along an optical fiber is described by two theories. Which theory is used to approximate the light acceptance and guiding properties of an optical fiber ?*

Ans. The ray theory.

Q. 22. *Meridional rays are classified as either bound or unbound rays. Bound rays propagate through the fiber according to what property?*

Ans. Total internal reflection.

■■■

3 OPTICAL FIBERS AND CABLES

Learning Objectives

Upon completion of this chapter, you should be able to do the following:

- Describe multimode and single mode step-index and graded-index fibers.
- Explain the terms refractive index profile, relative refractive index difference, and profile parameter.
- List the performance advantages of 62.5/125 μm multimode graded-index fibers.
- Identify the two basic types of single mode step-index fibers.
- Describe the vapor phase oxidation and direct-melt optical fiber fabrication procedures.
- Describe the fiber drawing process.
- List the benefits of cabled optical fibers over bare fibers.
- Identify the basic cable components, such as buffers, strength members, and jacket materials.
- Describe the material and design requirements imposed on military fiber optic cable designs.
- Describe the advantages and disadvantages of OFCC cable, stranded cable, and ribbon cable designs.

OPTICAL FIBER AND CABLE DESIGN

Optical fibers are thin cylindrical dielectric (non-conductive) waveguides used to send light energy for communication. Optical fibers consist of three parts: **the core, the cladding** and **the coating** or **buffer.** The choice of optical fiber materials and fiber design depends on operating conditions and intended application. Optical fibers are protected from the environment by incorporating the fiber into some type of cable structure. Cable strength members and outer jackets protect the fiber. Optical cable structure and material composition depend on the conditions of operation and the intended application.

OPTICAL FIBERS

Chapter 2 classified optical fibers as either single mode or multimode fibers. Fibers are classified according to the number of modes that they can propagate. Single mode fibers can propagate only the fundamental mode. Multimode fibers can propagate hundreds of modes. However, the classification of an optical fiber depends on more than the number of modes that a fiber can propagate.

An optical fiber's refractive index profile and core size further distinguish single mode and multimode fibers. The **refractive index profile** describes the value of refractive index as a function of radial distance at any fiber diameter. Fiber refractive index profiles classify single mode and multimode fibers as follows:

- Multimode step-index fibers
- Multimode graded-index fibers
- Single mode step-index fibers
- Single mode graded-index fibers.

In a **step-index** fiber, the refractive index of the core is uniform and undergoes an abrupt change at the core-cladding boundary. Step-index fibers obtain their name from this abrupt change called the step change in refractive index. In **graded-index** fibers, the refractive index of the core varies gradually as a function of radial distance from the fiber center.

Single mode and multimode fibers can have a step-index or graded-index refractive index profile. The performance of multimode graded-index fibers is usually superior to multimode step-index fibers. However, each type of multimode fiber can improve system design and operation depending on the intended application. Performance advantages for single mode graded-index fibers compared to single mode step-index fibers are relatively small. Therefore, single mode fiber production is almost exclusively step-index. Figure 3.1 shows

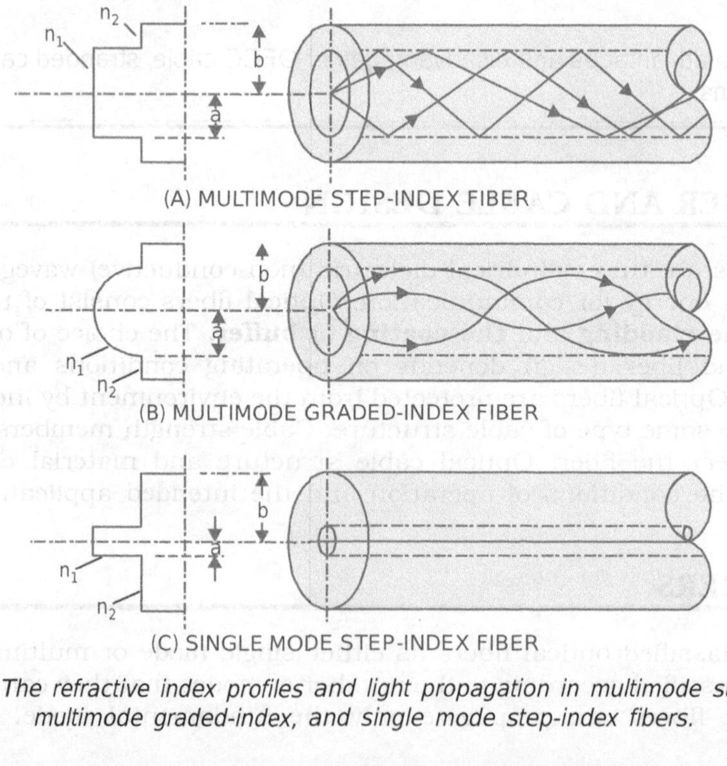

Fig. 3.1. *The refractive index profiles and light propagation in multimode step-index, multimode graded-index, and single mode step-index fibers.*

the refractive index profile for a multimode step-index fiber and a multimode graded-index fiber. Figure 3.1 also shows the refractive index profile for a single mode step-index fiber. Since light propagates differently in each fiber type. Figure 3.1 shows the propagation of light along each fiber.

In Chapter 2, you learned that fiber core size and material composition can affect system performance. A small change in core size and material composition affects fiber transmission properties, such as attenuation and dispersion.

When selecting an optical fiber, the system designer decides which fiber core size and material composition is appropriate.

Standard core sizes for multimode step-index fibers are 50 μm and 100 μm. Standard core sizes for multimode graded-index fibers are 50 μm, 62.5 μm, 85 μm and 100 μm. Standard core sizes for single mode fibers are between 8 μm and 10 μm. In most cases, the material used in the preparation of optical fibers is high-quality glass (SiO_2).

This glass contains very low amounts of impurities, such as water or elements other than silica and oxygen. Using high-quality glass produces fibers with low losses. Small amounts of some elements other than silica and oxygen are added to the glass material to change its index of refraction. These elements are called material dopants. Silica doped with various materials forms the refractive index profile of the fiber core and material dopants are discussed in more detail later in this chapter. Glass is not the only material used in fabrication of optical fibers. Plastics are also used for core and cladding materials in some applications.

A particular optical fiber design can improve fiber optic system performance.

Each single mode or multimode, step-index or graded-index, glass or plastic, or large or small core fiber has an intended application. The system designer must choose an appropriate fiber design that optimizes system performance in his application.

MULTIMODE STEP-INDEX FIBERS

A multimode step-index fiber has a core of radius (a) and a constant refractive index n_1. A cladding of slightly lower refractive index n_2 surrounds the core. Figure 3.2 shows the refractive index profile $n(r)$ for this type of fiber. $n(r)$ is equal to n_1 at radial distances $r < a$ (core). $n(r)$ is equal to n_2 at radial distances $r \geq a$ (cladding). Notice the step decrease in the value of refractive index at the core-cladding interface.

This step decrease occurs at a radius equal to distance (a). The difference in the core and cladding refractive index is the parameter Δ:

$$\Delta = \frac{n_1^2 - n_2^2}{2n_1^2}$$

Δ is the **relative refractive index difference.**

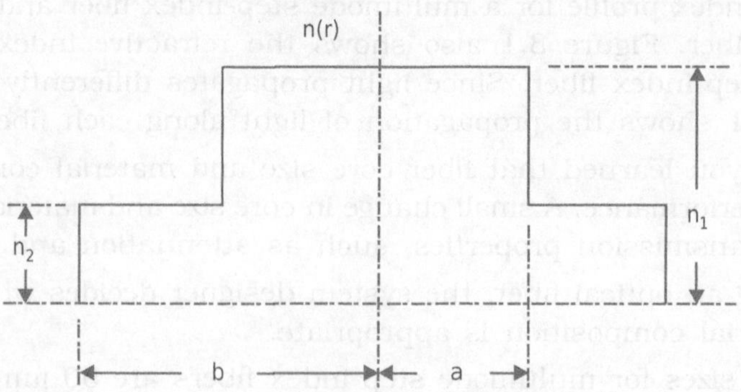

Fig. 3.2. *The refractive index profile for multimode step-index fibers.*

The ability of the fiber to accept optical energy from a light source is related to Δ. Δ also relates to the numerical aperture by

$$NA \simeq n_1 \sqrt{2\Delta}$$

The number of modes that multimode step-index fibers propagate depends on Δ and core radius (*a*) of the fiber. The number of propagating modes also depends on the wavelength (λ) of the transmitted light.

In a typical multimode step-index fiber, there are hundreds of propagating modes.

Most modes in multimode step-index fibers propagate far from cutoff.

Modes that are cutoff cease to be bound to the core of the fiber. Modes that are farther away from the cutoff wavelength concentrate most of their light energy into the fiber core. Modes that propagate close to cutoff have a greater percentage of their light energy propagate in the cladding. Since most modes propagate far from cutoff, the majority of light propagates in the fiber core. Therefore, in multimode step-index fibers, cladding properties, such as cladding diameter, have limited affect on mode (light) propagation.

Multimode step-index fibers have relatively large core diameters and large numerical apertures. A large core size and a large numerical aperture make it easier to couple light from a light-emitting diode (LED) into the fiber. Multimode step-index fiber core size is typically 50 μm or 100 μm.

Unfortunately, multimode step-index fibers have limited bandwidth capabilities.

Dispersion, mainly modal dispersion, limits the bandwidth or information-carrying capacity of the fiber. System designers consider each factor when selecting an appropriate fiber for each particular application.

Multimode step-index fiber selection depends on system application and design. Short-haul, limited bandwidth, low-cost applications typically use multimode step-index fibers.

MULTIMODE GRADED-INDEX FIBERS

A multimode graded-index fiber has a core of radius (*a*). Unlike step-index fibers, the value of the refractive index of the core (n_1) varies according to the radial distance (r). The value of n_1 decreases as the distance (r) from the center of the fiber increases.

The value of n_1 decreases until it approaches the value of the refractive index of the cladding (n_2). The value of n_1 must be higher than the value of n_2 to allow for proper mode propagation. Like the step-index fiber, the value of n_2 is constant and has a slightly lower value than the maximum value of n_1. The relative refractive index difference (Δ) is determined using the maximum value of n_1 and the value of n_2.

Figure 3.3 shows a possible refractive index profile $n(r)$ for a multimode graded-index fiber. Notice the parabolic refractive index profile of the core. The **profile parameter** (α) determines the shape of the core's profile. As the value of α increases, the shape of the core's profile changes from a triangular shape to step as shown in Fig. 3.4. Most multimode graded-index fibers have a parabolic refractive index profile. Multimode fibers with near parabolic graded-index profiles provide the best performance. Unless otherwise specified, when discussing multimode graded-index fibers, assume that the core's refractive index profile is parabolic ($\alpha = 2$).

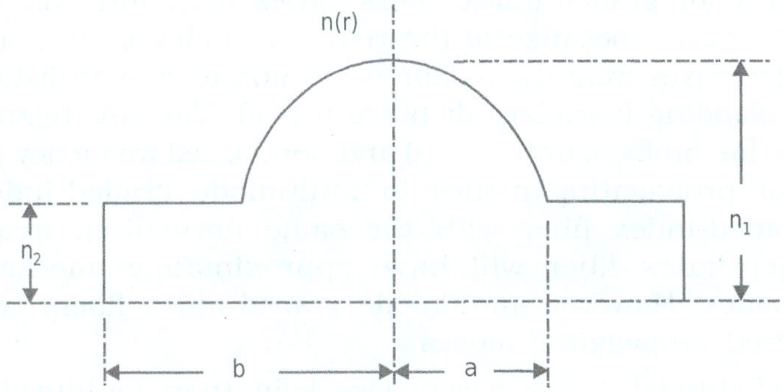

Fig. 3.3. *The refractive index profile for multimode graded-index fibers.*

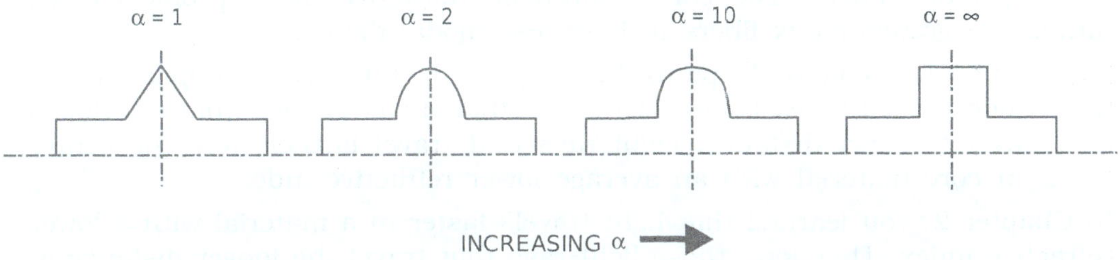

Fig. 3.4. *The refractive index profiles for different values of* α.

Light propagates in multimode graded-index fibers according to refraction and total internal reflection. The gradual decrease in the core's refractive index from the center of the fiber causes the light rays to be refracted many times. The light rays become refracted or curved, which increases the angle of incidence at the next point of refraction. Total internal reflection occurs when the angle of incidence becomes larger than the critical angle of incidence. Figure 3.5 shows the process of refraction and total internal reflection of light in multimode graded-index fibers. Figure 3.5 also illustrates the boundaries of different values of core refractive index by dotted lines. Light rays may be reflected to the axis of the fiber before reaching the core-cladding interface.

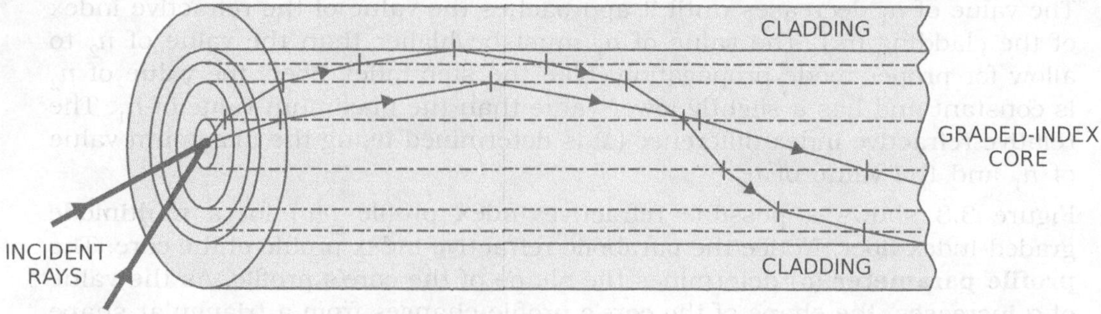

Fig. 3.5. *Refractive index grading and light propagation in multimode graded-index fibers.*

The NA of a multimode graded-index fiber is at its maximum value at the fiber axis. This NA is the **axial numerical aperture [NA(0)]. NA(0) is approximately equal to** $n_1\sqrt{2\Delta}$.

However, the NA for graded-index fibers varies as a function of the radial distance (r). NA varies because of the refractive index grading in the fiber's core. The NA decreases from the maximum, NA(0), to zero at distances greater than the core-cladding boundary distance ($r > a$). The NA, relative refractive index difference (Δ), profile parameter (α) and normalized frequency (V) determine the number of propagating modes in multimode graded-index fibers. A multimode graded-index fiber with the same normalized frequency as a multimode step-index fiber will have approximately one-half as many propagating modes. However, multimode graded-index fibers typically have over one-hundred propagating modes.

Multimode graded-index fibers accept less light than multimode step-index fibers with the same core Δ. However, graded-index fibers usually outperform the step-index fibers. The core's parabolic refractive index profile causes multimode graded-index fibers to have less modal dispersion.

Figure 3.5 shows possible paths that light may take when propagating in multimode graded-index fibers. Light rays that travel farther from the fiber's axis travel a longer distance. Light rays that travel farther from the center travel in core material with an average lower refractive index.

In Chapter 2, you learned that light travels faster in a material with a lower refractive index. Therefore, those light rays that travel the longer distance in the lower refractive index parts of the core travel at a greater average velocity. This means that the rays that travel farther from the fiber's axis will arrive at each point along the fiber at nearly the same time as the rays that travel close to the fiber's axis. The decrease in time difference between light rays reduces modal dispersion and increases multimode graded-index fiber bandwidth. The increased bandwidth allows the use of multimode graded-index fibers in most applications.

Most present day applications that use multimode fiber use graded-index fibers. The basic design parameters are the fiber's core and cladding size and Δ. Standard multimode graded-index fiber core and cladding sizes are 50/125 μm, 62.5/125 μm, 85/125 μm and 100/140 μm. Each fiber design has a specific Δ that improves fiber performance. Typical values of Δ are around 0.01 to 0.02. Although no single multimode graded-index fiber design is appropriate for all applications, the 62.5/125 μm fiber with a Δ of 0.02 offers the best overall performance.

A multimode graded-index fiber's source-to-fiber coupling efficiency and insensitivity to microbending and macrobending losses are its most distinguishing characteristics. The fiber core size and Δ affect the amount of power coupled into the core and loss caused by microbending and macrobending. Coupled power increases with both core diameter and Δ, while bending losses increase directly with core diameter and inversely with Δ. However, while these values favour high Δ, a smaller Δ improves fiber bandwidth.

In most applications, a multimode graded-index fiber with a core and cladding size of 62.5/125 μm offers the best combination of the following properties:

- Relatively high source-to-fiber coupling efficiency
- Low loss
- Low sensitivity to microbending and macrobending
- High bandwidth
- Expansion capability.

For example, local area network (LAN) and shipboard applications use multimode graded-index fibers with a core and cladding size of 62.5/125 μm. In LAN-type environments, macrobend and microbend losses are hard to predict. Cable tension, bends, and local tie-downs increase macrobend and microbend losses. In shipboard applications, a ship's cable-way may place physical restrictions, such as tight bends, on the fiber during cable plant installation. The good microbend and macrobend performance of 62.5/125 Δm fiber permits installation of a rugged and robust cable plant. 62.5/125 Δm multimode graded-index fibers allow for uncomplicated growth because of high fiber bandwidth capabilities for the expected short cable runs on board ships.

SINGLE MODE STEP-INDEX FIBERS

There are two basic types of single mode step-index fibers: matched clad and depressed clad. **Matched cladding** means that the fiber cladding consists of a single homogeneous layer of dielectric material. **Depressed cladding** means that the fiber cladding consists of two regions: the inner and outer cladding regions.

Matched-clad and depressed-clad single mode step-index fibers have unique refractive index profiles.

A matched-clad single mode step-index fiber has a core of radius (*a*) and a constant refractive index n_1. A cladding of slightly lower refractive index surrounds the core. The cladding has a refractive index n_2. Figure 3.6 shows the refractive index profile $n(r)$ for the matched-clad single mode fiber.

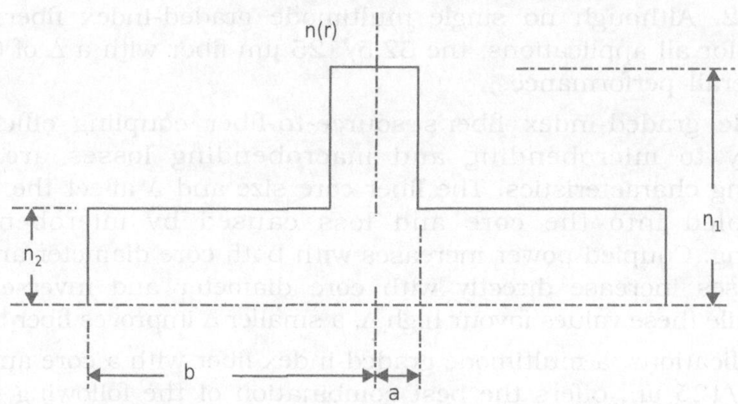

Fig. 3.6. *Matched-clad refractive index profile.*

Figure 3.7 shows the refractive index profile $n(r)$ for the depressed-clad single mode fiber. A depressed-clad single mode step-index fiber has a core of radius (a) with a constant refractive index n_1. A cladding, made of two regions, surrounds the core.

An inner cladding region surrounds the core of the fiber and has a refractive index of n_2. The inner cladding refractive index n_2 is lower than the core's refractive index n_1.

An outer cladding region surrounds the inner cladding region and has a higher refractive index n_3 than the inner cladding region. However, the outer cladding refractive index n_3 is lower than the core's refractive index n_1.

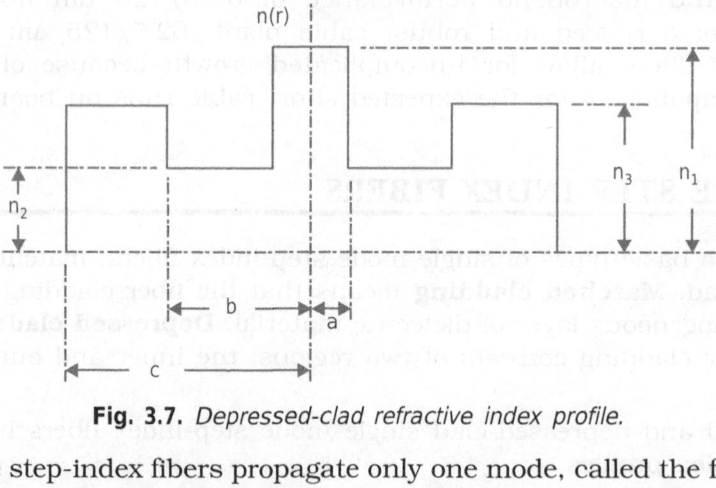

Fig. 3.7. *Depressed-clad refractive index profile.*

Single mode step-index fibers propagate only one mode, called the fundamental mode. Single mode operation occurs when the value of the fiber's normalized frequency is between 0 and 2.405 ($0 \leq V < 2.405$). The value of V should remain near the 2.405 level. When the value of V is less than 1, single mode fibers carry a majority of the light power in the cladding material. The portion of light transmitted by the cladding material easily radiates out of the fiber. For example, light radiates out of the cladding material at fiber bends and splices.

Single mode fiber cutoff wavelength is the smallest operating wavelength when single mode fibers propagate only the fundamental mode. At this wavelength, the 2nd-order mode becomes lossy and radiates out of the fiber core. As the operating wavelength becomes longer than the cutoff wavelength, the fundamental mode becomes increasingly lossy.

The higher the operating wavelength is above the cutoff wavelength, the more power is transmitted through the fiber cladding. As the fundamental mode extends into the cladding material, it becomes increasingly sensitive to bending loss.

Single mode fiber designs include claddings of sufficient thickness with low absorption and scattering properties to reduce attenuation of the fundamental mode. To increase performance and reduce losses caused by fiber bending and splicing, fiber manu facturers adjust the value of V. To adjust the value of V, they vary the core and cladding sizes and relative refractive index difference (Δ).

A single mode step-index fiber has low attenuation and high bandwidth properties. Present applications for single mode fibers include long-haul, high-speed telecommunication systems. Future applications include single mode fibers for sensor systems. However, the current state of single mode technology makes installation of single mode systems expensive and difficult. Short cable runs, low to moderate bandwidth requirements, and high component cost make installation of single mode fiber shipboard systems impractical at this time.

SINGLE MODE GRADED-INDEX FIBERS

There are several types of single mode graded-index fibers. These fibers are not standard fibers and are typically only used in specialty applications.

FIBER ALTERNATIVES

In most applications, the standard multimode and step-index single mode optical fibers mentioned before have significant performance advantages over conventional copper-based systems. However, performance requirements and cost restraints may prohibit the use of these fibers in certain applications.

Fiber manufacturers modify standard multimode and single mode fiber material composition and structural design to meet these additional requirements. Optical fiber design can depart from a traditional circular core and cladding, low-loss glass design. The intent of each change is to increase performance and reduce cost.

Optical fibers composed of plastic have been in use longer than glass fibers. Types of standard fibers using plastics include multimode step-index and graded-index fibers. Multimode step-index and graded-index **plastic clad silica** (PCS) fibers exist. PCS fibers have a silica glass core and a plastic cladding. Normally, PCS fibers are cheaper than all-glass fibers but have limited performance characteristics. PCS fibers lose more light through a plastic cladding than a glass cladding.

Multimode step-index fibers may also have a plastic core and cladding. All-plastic fibers have a higher NA, a larger core size, and cost less to manufacture. However, all-plastic fibers exhibit high loss in the thousands of decibels per kilometer. This high loss is caused by impurities and intrinsic absorption. PCS and all-plastic fibers are used in applications typically characterized by one or all of the following:

- High NA
- Low bandwidth
- Tight bend radius
- Short length (less than 10 m to 20 m)
- Low cost.

Improved fabrication techniques provide the opportunity to experiment with material composition in both multimode and single mode fibers. Fiber manufacturers fabricate optical fibers using glass material whose characteristics improve system performance in the far infrared region. Fiber manufacturers add dopant material to reduce fiber loss and limit material and structural imperfections. Fiber material used in fabrication of low-loss, long wavelength optical fibers include the following:

- Heavy-metal fluorides (such as zirconium and beryllium fluoride)
- Chalcogenide glasses (such as arsenic/sulphur)
- Crystalline materials (such as silver bromide and silver chloride).

In shipboard applications, stringent environmental requirements dictate the design of special optical fibers. In some cases, manufacturers hermetically coat optical fibers to increase survivability and reliability in high-moisture and high-strain environments. Manufacturers also design radiation-hard fibers for nuclear power, space, and military systems. Radiation resistant fibers operate after exposure to nuclear radiation. Shipboard system performance requirements determine whether the use of hermetic and radiation resistant fibers or less costly commercial optical fibers is necessary.

FABRICATION OF OPTICAL FIBERS

Basically, fiber manufacturers use two methods to fabricate multimode and single mode glass fibers. One method is vapor phase oxidation, and the other method is direct-melt process. In **vapor phase oxidation,** gaseous metal halide compounds, dopant material, and oxygen are oxidized (burned) to form a white silica powder (SiO_2). Manufacturers call SiO_2 the **soot.**

Manufacturers deposit the soot on the surface of a glass substrate (mandrel) or inside a hollow tube by one of the following three methods:

1. Outside Vapor Phase Oxidation (OVPO)

- Inside Vapor Phase Oxidation (IVPO)
- Vapor Phase Axial Deposition (VAD).

The soot forms the core and cladding material of the preform. The refractive index of each layer of soot is changed by varying the amount of dopant material being oxidized. Figures 3.8, 3.9 and 3.10 illustrate the different vapor phase oxidation preform preparation methods.

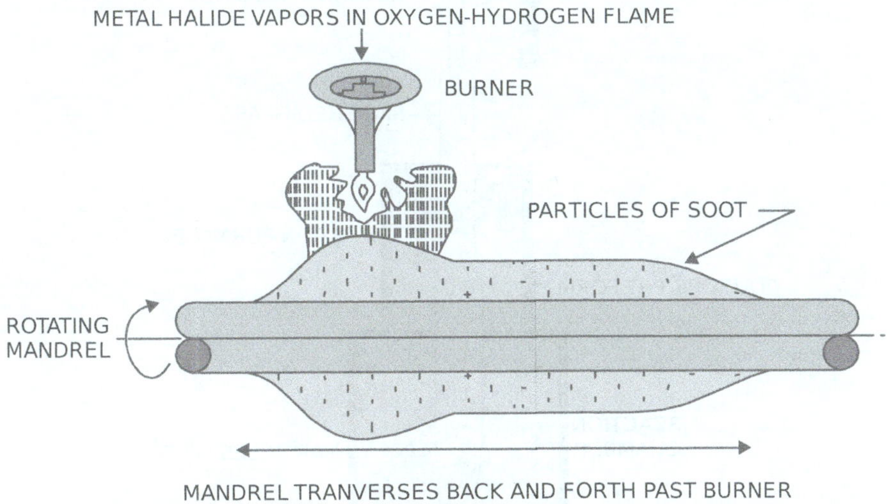

Fig. 3.8. *OVPO preform preparation.*

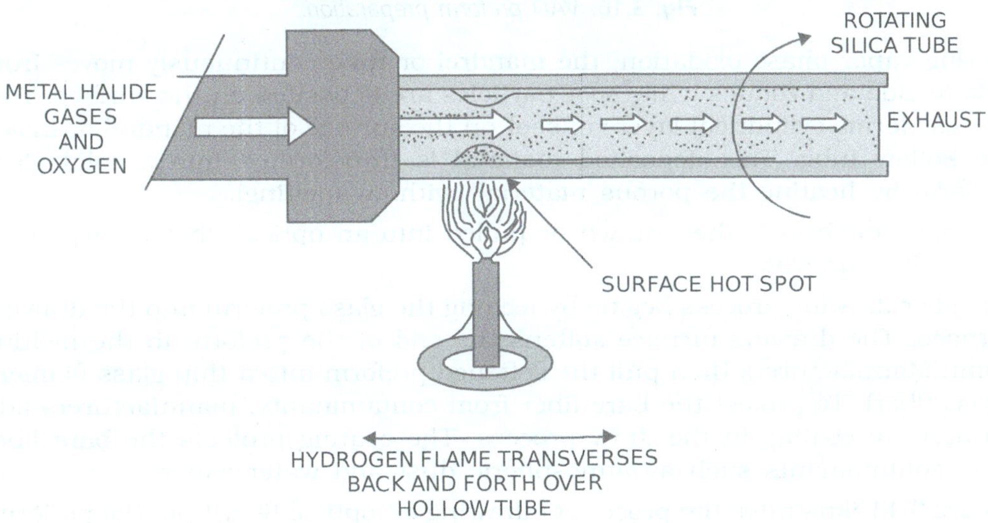

Fig. 3.9. *IVPO preform preparation.*

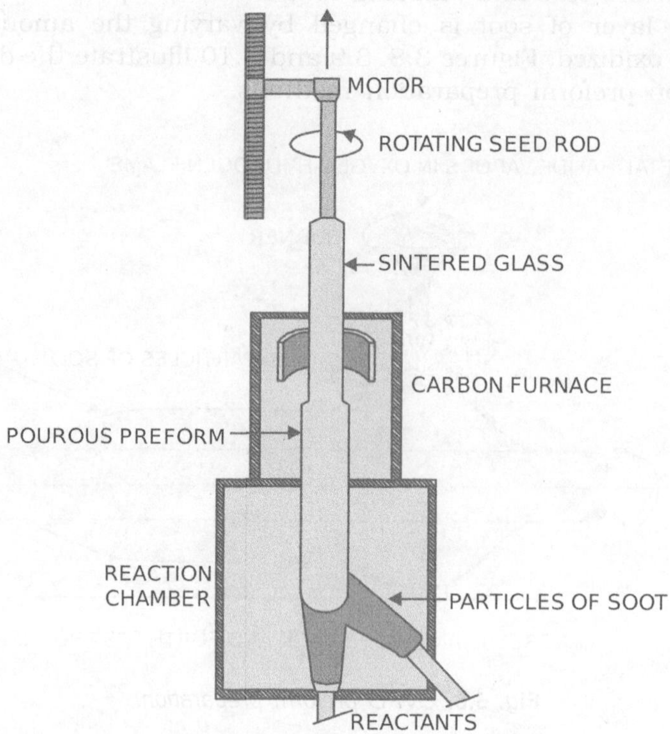

Fig. 3.10. *VAD preform preparation.*

During vapor phase oxidation, the mandrel or tube continuously moves from side to side and rotates while soot particles are deposited on the surface. This process forms cylindrical layers of soot on the surface of the mandrel or inside the hollow tube. This deposited material is transformed into a solid glass preform by heating the porous material (without melting).

The solid preform is then drawn or pulled into an optical fiber by a process called fiber drawing.

The fiber drawing process begins by feeding the glass preform into the drawing furnace. The drawing furnace softens the end of the preform to the melting point. Manufacturers then pull the softened preform into a thin glass filament (glass fiber). To protect the bare fiber from contaminants, manufacturers add an acrylate coating in the draw process. The coating protects the bare fiber from contaminants such as atmospheric dust and water vapor.

Figure 3.11 illustrates the process of drawing an optical fiber from the preform.

In the **direct-melt process,** multicomponent glass rods form the fiber structure. Rods of multicomponent glass combine in a molten state to form the fiber core and cladding. The double-crucible method is the most common direct-melt process. The double-crucible method combines the molten rods into a single preform using two concentric crucibles.

Optical fibers are drawn from this molten glass using a similar fiber drawing process as in vapor phase oxidation. Figure 3.12 illustrates the double-crucible drawing process.

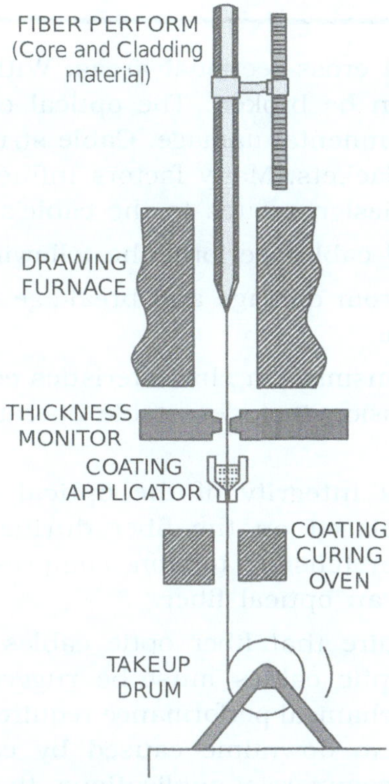

Fig. 3.11. *Fiber drawing process.*

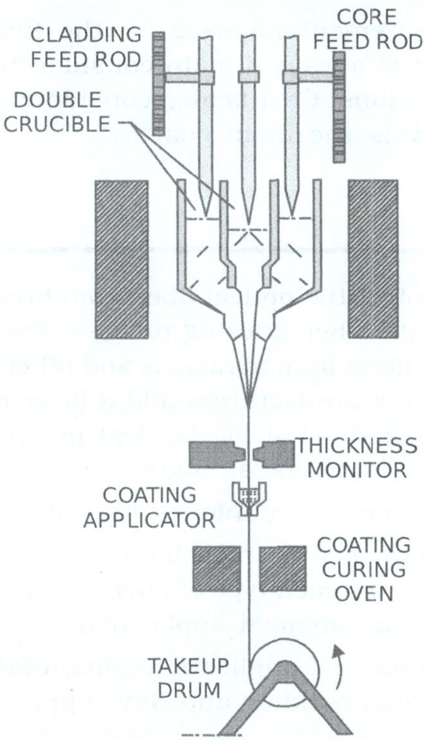

Fig. 3.12. *Double-crucible fiber drawing process.*

OPTICAL CABLES

Optical fibers have small cross-sectional areas. Without protection, optical fibers are fragile and can be broken. The optical cable structure protects optical fibers from environmental damage. Cable structure includes buffers, strength members, and jackets. Many factors influence the design of fiber optic cables. The cable design relates to the cable's intended application.

Properly designed optical cables perform the following functions:

- Protect optical fibers from damage and breakage during installation and over the fiber's lifetime.
- Provide stable fiber transmission characteristics compared with uncabled fibers. Stable transmission includes stable operation in extreme climate conditions.
- Maintain the physical integrity of the optical fiber by reducing the mechanical stresses placed on the fiber during installation and use. Static fatigue caused by tension, torsion, compression, and bending can reduce the lifetime of an optical fiber.
- Navy applications require that fiber optic cables meet stringent design specifications. Fiber optic cables must be rugged to meet the optical, environmental, and mechanical performance requirements imposed by Navy systems. Critical system downtime caused by cable failure cannot be tolerated. However, in commercial applications, the requirements imposed on cable designs are not as stringent.
- Each additional requirement imposed on the fiber optic cable design adds to its cost. Cost is always a main consideration of cable designers in commercial applications. Cost is also considered in Navy applications, but system reliability is the main goal.

FIBER BUFFERS

Coatings and buffers protect the optical fiber from breakage and loss caused by microbends. During the fiber drawing process, the addition of a primary coating protects the bare glass from abrasions and other surface contaminants. For additional protection, manufacturers add a layer of buffer material. The buffer material provides additional mechanical protection for the fiber and helps preserve the fiber's inherent strength.

Manufacturers use a variety of techniques to buffer optical fibers.

The types of fiber buffers include tight-buffered, loose-tube, and gel-filled loose-tube. Figure 3.13 shows each type of fiber buffer. The choice of buffering techniques depends on the intended application.

In large fiber count commercial applications, manufacturers use the loose-tube buffers. In commercial building and Navy applications, manufacturers use tight buffers.

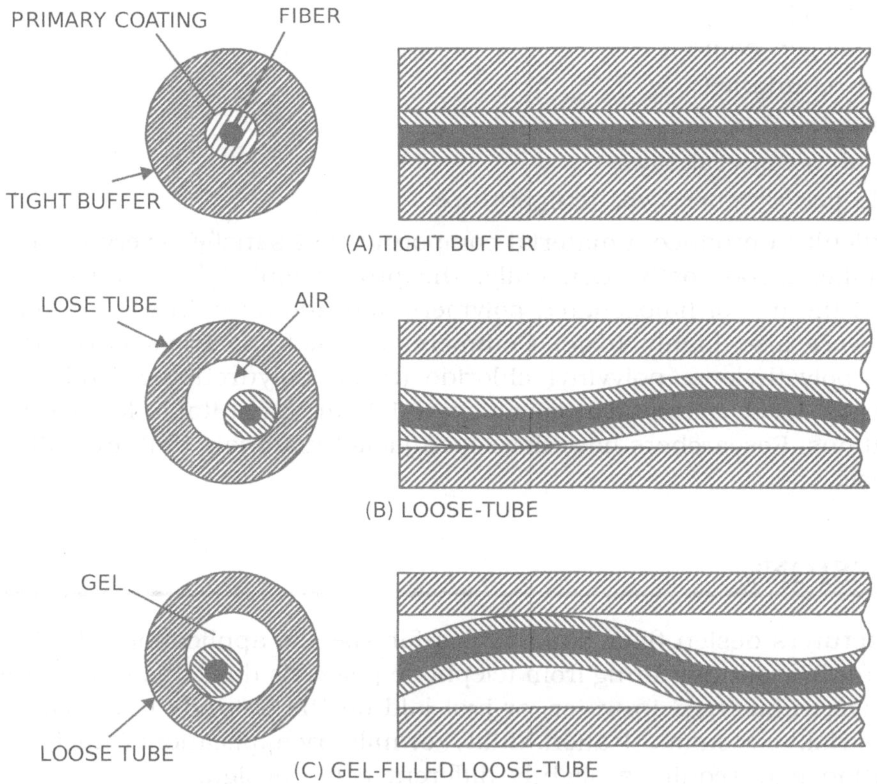

Fig. 3.13. *Tight-buffered, loose-tube and gel-filled loose-tube buffer techniques.*

CABLE STRENGTH AND SUPPORT MEMBERS

Fiber optic cables use strength members to increase the cables' strength and protect the fiber from strain. Fiber optic cables may use central support members in cable construction. The central support members generally have buffered fibers or single fiber sub-cables stranded over their surface in a structured, helical manner. The central members may support the optical fibers as cable strength members or may only serve as fillers. Strength and support members must be light and flexible. In commercial applications, the materials used for strength and support include steel wire and textile fibers (such as nylon and arimid yarn). They also include carbon fibers, glass fibers, and glass reinforced plastics. For Navy applications, only non-metallic strength and support members are allowed.

CABLE JACKET OR SHEATH MATERIAL

The jacket or sheath material provides extra environmental and mechanical protection. Jacket materials for Navy cables have the following properties:

- Low smoke generation

- Low toxicity
- Low halogen content
- Flame retardance
- Fluid resistance
- High abrasion resistance
- Stable performance over temperature.

It is difficult to produce a material compound that satisfies every requirement without being too costly. Originally, the production of fire retardant cables included the use of halogenated polymers and additives. These fire retardant cables were also highly toxic. Commercial jacket materials currently used include polyethylene, polyvinyl chloride (PVC), polyurethane, and polyester elastomers. Most commercial jacket materials are unsuitable for use in Navy applications. Researchers have developed jacket materials that are suitable for Navy use.

CABLE DESIGNS

Manufacturers design fiber optic cables for specific applications. Is the cable buried underground or hung from telephone poles? Is the cable snaked through cableways, submerged in water, or just laid on the ground? Is the cable used in industrial, telecommunication, utility, or military applications? Each different application may require a slightly different cable design.

Agreement on standard cable designs is difficult. Cable design choices include jacket materials, water blocking techniques, and the number of fibers to place within the cable. The cable design chosen depends on the cable's intended application. There are presently many types of fiber optic cables. Some fiber optic cables are used in commercial applications, while others are used in military applications. Standard commercial cable designs will develop over time as fiber optic technology becomes more established. However, this chapter provides only a short discussion on cable designs considered for Navy applications.

Navy systems require that fiber optic cables meet stringent environmental conditions. The types of cable designs considered by the Navy include the optical fiber cable component (OFCC), stranded, and ribbon cable designs.

The cable must meet minimal levels of performance in safety (low smoke, low toxicity, low halogen content, etc.), durability (able to withstand shock, vibration, fluids, etc.), and optical performance. The cable must also be easy to install and repair. These factors greatly influence the design of the cables.

Optical Fiber Cable Component (OFCC) Cable

An OFCC cable consists of individual single fiber cables, called **optical fiber cable components (OFCCs).** OFCCs are a tight-buffered fiber surrounded by arimid yarn and a low-halogen outer jacket. The OFCC outer diameter is typically 2 millimeters (mm). The fiber is typically buffered with a polyester elastomer to a total diameter of 900 μm. Figure 3.14 illustrates the design of the OFCCs. The size of the OFCCs limits the amount of fibers contained within an OFCC cable. An OFCC cable generally contains less than 36 fibers (OFCCs). An OFCC cable of 0.5 inch cable outer diameter can accommodate about 12 fibers.

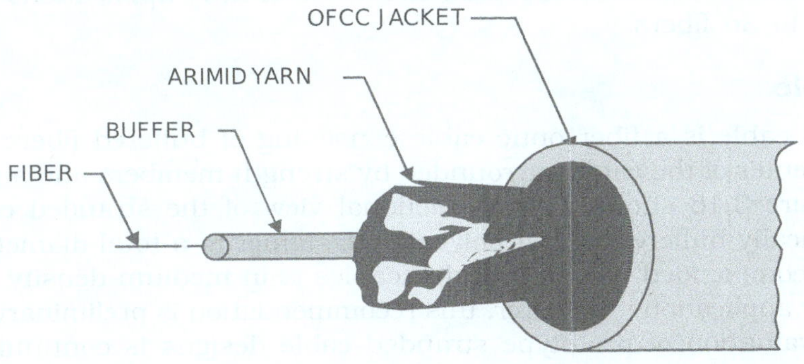

Fig. 3.14. *The design of optical fiber cable components (OFCCs).*

Figure 3.15 shows an isometric view of a four-fiber shipboard OFCC cable. In this multifiber cable design, the OFCCs surround a flexible central member in a helical manner. The central member may add to cable strength or only support the OFCCs. For additional protection, two layers of arimid yarn strength members encase the OFCC units. These strength members are stranded in opposing lays to minimize microbending of the fibers.

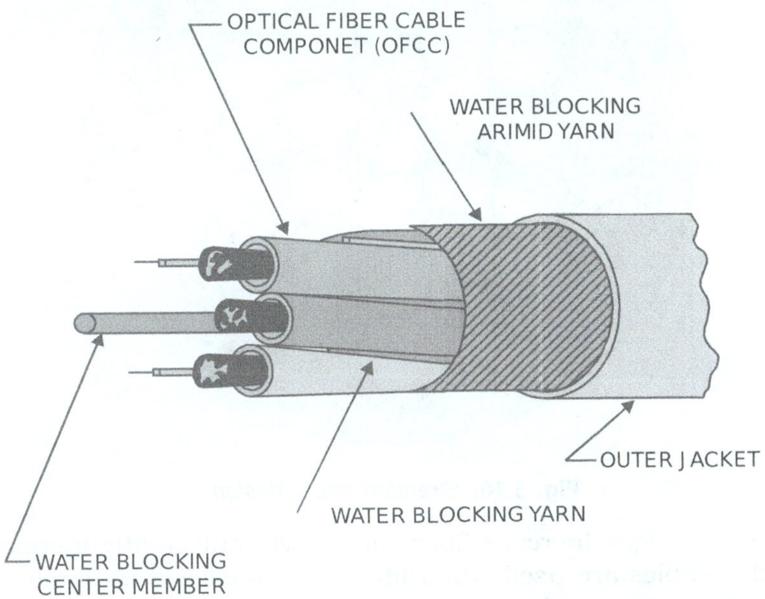

Fig. 3.15. *An isometric view of a four-fiber shipboard OFCC cable.*

The arimid yarn strength members may be treated with polymers that are water absorbing, blocking, and sealing. This treatment eliminates the need for additional water blocking protection. Finally, a low-halogen, flame-resistant outer jacket is extruded over the strength members.

OFCC cables are easy to handle because each cable contains its own subcable, the OFCC. These OFCC subcables make it easy to reconfigure systems and handle individual fibers. Rugged OFCC cable design permits cable use in harsh environments, including Navy applications. OFCC-type cable is recommended for use in low-density (less than 24 fibers) Navy applications. OFCC-type cable is also being evaluated for use in Navy applications with fiber counts up to 36 fibers.

Stranded Cable

A stranded cable is a fiber optic cable consisting of buffered fibers stranded down the center of the cable surrounded by strength members and a protective jacket. Figure 3.16 shows a cross-sectional view of the stranded cable. The fiber is typically buffered with a polyester elastomer to a total diameter of 900 μm. The recommended use of stranded cables is in medium-density (24 to 72 fibers) Navy applications. However, this recommendation is preliminary. Further test and evaluation of prototype stranded cable designs is continuing. Final approval of the stranded cable will occur only after prototype cables have passed all tests.

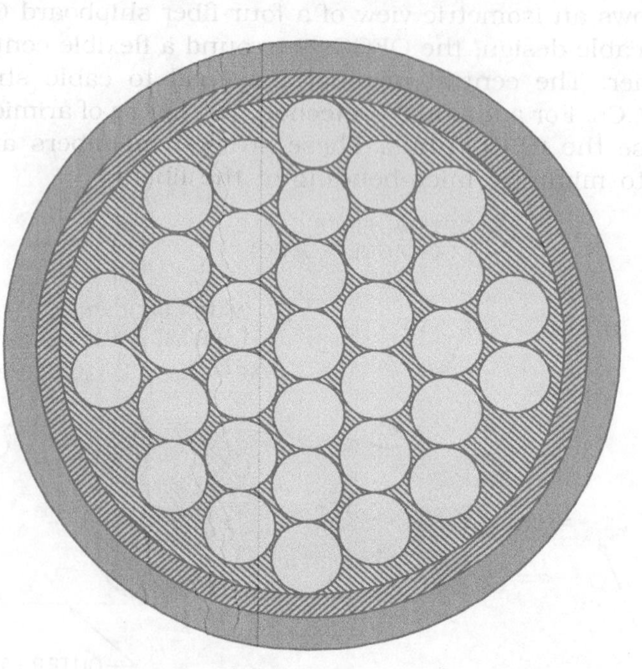

Fig. 3.16. *Stranded cable design.*

Stranded cable designs increase fiber counts without greatly increasing cable size. Stranded cables are used when fiber counts exceed the limits of OFCC-type cables. For example, the stranded cable design can accommodate about 48 fibers in a 0.5 inch cable. The OFCC cable design can accommodate around 12 fibers. The individual fiber is not protected as well in the stranded design as in the OFCC design. For this reason more care is required in handling the individual fibers in the stranded design. The primary problem of the stranded cable design is in meeting the waterblocking requirements. Once manufacturers correct this design problem, the Navy expects that the stranded cable design will meet Navy performance requirements.

Ribbon Cable

A ribbon cable consists of optical fiber ribbons stranded down the center of the cable surrounded by a protective tube, strength members, and an outer jacket. The fiber optic ribbon consists of multiple-coated, 250 μm diameter fibers sandwiched in a plastic material. Figure 3.17 shows a cross-sectional

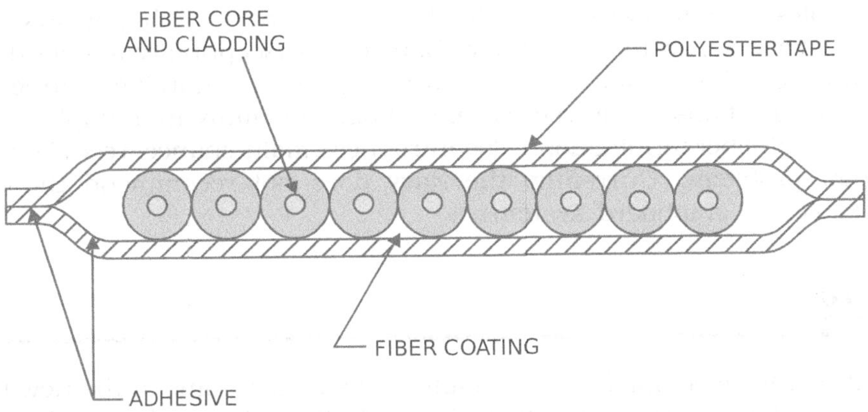

Fig. 3.17. *Cross-section of a fiber optic ribbon.*

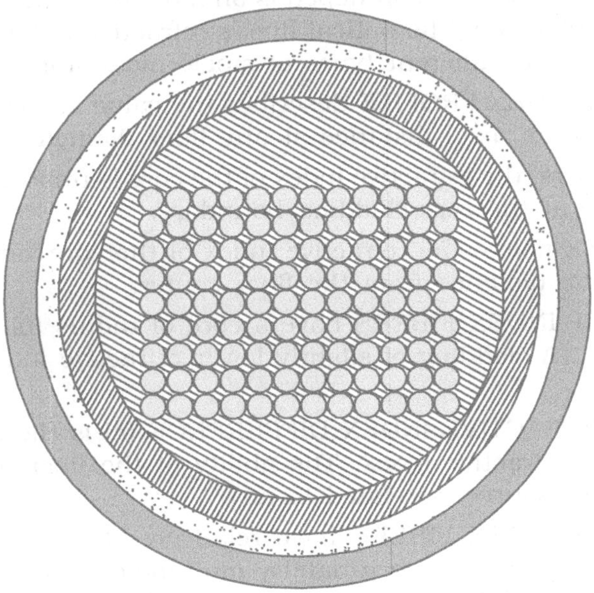

Fig. 3.18. *Ribbon cable cross-sectional array of fibers.*

view of a 12-fiber ribbon. Cable manufacturers stack these ribbons to form a rectangular cross-sectional array of fibers. Stacked ribbons are the basic building blocks of the ribbon cable. Figure 3.18 illustrates this cross-sectional array of ribbons. Manufacturers introduce a controlled twist to the stacked ribbons to minimize fiber stress when the cable is bent. An inner plastic tube, strength members, and an outer protective jacket surround the stacked ribbons, providing environmental protection.

The ribbon cable design has the highest fiber capacity. Ribbon cables can hold 204 fibers in a 0.5-inch cable. However, ribbon cables have worse bend performance than OFCC and stranded cables. Ribbon cables also have the poorest waterblocking capabilities of the three cable designs. The bend performance is expected to worsen if manufacturers add appropriate compounds to increase waterblocking capabilities.

Ribbon cables are also hard to handle. Individual fibers are highly susceptible to damage when separated from the ribbon. This susceptibility to fiber damage during fiber breakout makes it necessary to perform multifiber connections. Multifiber connections can introduce single points of failure in multiple systems. The use of multifiber terminations also introduces maintenance, reconfiguration, and repair problems. Currently, the Navy does not recommend the use of ribbon cables in shipboard systems.

SUMMARY

Now that you have completed this chapter, let's review some of the new terms, concepts, and ideas that you have learned. You should have a thorough understanding of these principles before moving on to Chapter 4.

OPTICAL FIBER CLASSIFICATION depends on more than the number of modes that a fiber can propagate. The optical fiber's refractive index profile and core size further distinguish different types of single mode and multimode fibers.

The REFRACTIVE INDEX PROFILE describes the value of the fiber's refractive index as a function of axial distance at any fiber diameter.

In STEP-INDEX fibers, the refractive index of the core is uniform and undergoes an abrupt change at the core-cladding boundary.

In GRADED-INDEX fibers, the refractive index of the core varies gradually as a function of radial distance from the fiber center.

MULTIMODE STEP-INDEX FIBERS have a core of radius (a), and a constant refractive index n_1. A cladding of slightly lower refractive index n_2 surrounds the core.

The RELATIVE REFRACTIVE INDEX DIFFERENCE (Δ) is the difference in the core and cladding refractive index. The ability of the fiber to accept optical energy from a light source is related to Δ.

MULTIMODE STEP-INDEX FIBERS have relatively large core diameters and large numerical apertures. Unfortunately, multimode step-index fibers have limited bandwidth capabilities and poor bend performance.

Short-haul, limited bandwidth, low-cost applications use multimode step-index fibers.

MULTIMODE GRADED-INDEX FIBERS have a core of radius (a). Unlike step-index fibers, the value of the refractive index of the core (n_1) varies according to the radial distance (r). The value of n_1 decreases until it approaches the value of the refractive index of the cladding (n_2). Like the step-index fiber, the value of n_2 is constant and has a slightly lower refractive index than n_1.

The PROFILE PARAMETER (α) determines the shape of the core's refractive index profile. As the value of α; increases, the shape of the core's profile changes from a triangular shape to a step.

The gradual decrease in the core's refractive index from the center of the fiber causes propagating modes to be refracted many times.

Multimode graded-index fibers have less MODAL DISPERSION than multimode step-index fibers. Lower modal dispersion means that multimode graded-index fibers have higher bandwidth capabilities than multimode step-index fibers.

SOURCE-TO-FIBER COUPLING EFFICIENCY AND INSENSITIVITY TO MICROBENDING AND MACROBENDING LOSSES are distinguishing characteristics of multimode graded-index fibers. 62.5 μm fibers offer the best overall performance for multimode graded-index fibers.

Coupled power increases with both core diameter and Δ, while bending losses increase directly with core diameter and inversely with Δ. However, a smaller Δ improves fiber bandwidth.

MATCHED-CLAD and DEPRESSED-CLAD are two types of single mode step-index fibers. Matched cladding means that the fiber cladding is a single homogeneous layer of dielectric material. Depressed cladding means that the fiber cladding consists of two regions: an **inner** and **outer** cladding region.

SINGLE MODE FIBER CUTOFF WAVELENGTH is the smallest operating wavelength where single mode fibers propagate only the fundamental mode. At this wavelength, the 2nd-order mode becomes lossy and radiates out of the fiber core.

SINGLE MODE FIBERS have low attenuation and high-bandwidth properties. Present applications for single mode fibers include long-haul, high-speed telecommunication systems.

VAPOR PHASE OXIDATION and DIRECT-MELT PROCESS are two methods of fabricating multimode and single mode optical fibers.

CABLE STRUCTURES include buffers, strength members, and the jacket, or sheath.

TIGHT-BUFFERED, LOOSE-TUBE, and GEL-FILLED LOOSE-TUBE are types of fiber optic buffering techniques.

FIBER OPTIC CABLES use strength members to increase the cable's strength and protect the optical fibers from strain.

JACKET MATERIAL should have low smoke generation, low toxicity, low-halogen content, flame retardance, fluid resistance, high abrasion resistance, and stable performance over temperature.

Navy systems require that fiber optic cables meet stringent environmental conditions. The types of cable designs considered by the Navy include the OPTICAL FIBER CABLE COMPONENT (OFCC), STRANDED, and RIBBON cable designs.

REVIEW QUESTIONS AND ANSWERS

Q. 1. *Describe the term "refractive index profile."*

Ans. Refractive index profile describes the value of refractive index as a function of radial distance at any fiber diameter.

Q. 2. *The refractive index of a fiber core is uniform and undergoes an abrupt change at the core-cladding boundary. Is this fiber a step-index or graded-index fiber?*

Ans. Step-index.

Q. 3. *Multimode optical fibers can have a step-index or graded-index refractive index profile. Which fiber, multimode step-index or multimode graded-index fiber, usually performs better?*

Ans. Multimode graded-index fiber.

Q. 4. *List the standard core sizes for multimode step-index, multimode graded-index, and single mode fibers.*

Ans. Multimode step-index fibers: 50 μm and 100 μm. Multimode graded-index fibers: 50 μm, 62.5 μm, 85 μm, and 100 μm. Single mode fibers: between 8 μm and 10 μm.

Q. 5. *Multimode step-index fibers have a core and cladding of constant refractive index n_1 and n_2, respectively. Which refractive index, the core or cladding, is lower?*

Ans. Cladding.

Q. 6. *In multimode step-index fibers, the majority of light propagates in the fiber core for what reason?*

Ans. Most modes in multimode step-index fibers propagate far from cutoff.

Q. 7. *Multimode step-index fibers have relatively large core diameters and large numerical apertures. These provide what benefit?*

Ans. Make it easier to couple light from a light-emitting diode (LED) into the fiber.

Q. 8. *The profile parameter (α) determines the shape of the multimode graded-index core's refractive index profile. As the value of the α increases, how does the core's profile change?*

Ans. From a triangular shape to step.

Q. 9. *Light propagates in multimode graded-index fibers according to refraction and total internal reflection. When does total internal reflection occur?*

Ans. When the angle of incidence becomes larger than the critical angle of incidence.

Q. 10. *What four fiber properties determine the number of modes propagating in a multimode graded-index fiber?*

Ans. Numerical aperture (NA), relative refractive index difference (α), profile parameter (α) and normalized frequency (V).

Q. 11. *Light travels faster in a material with a lower refractive index. Therefore, light rays that travel a longer distance in a lower refractive index travel at a greater average velocity. What effect does this have on multimode graded-index fiber modal dispersion and bandwidth?*

Ans. Decreases the time difference between light rays, which reduces modal dispersion and increases fiber bandwidth.

Q. 12. *What multimode graded-index fiber offers the best overall performance for most applications?*

Ans. 62.5/125 μm multimode graded-index fiber.

Q. 13. *What are the most distinguishing characteristics of a multimode graded-index fiber?*

Ans. Source-to-fiber coupling efficiency and insensitivity to microbending and macrobending losses.

Q. 14. *How are source-to-fiber coupling and microbending and macrobending losses affected by changes in core diameter and Δ?*

Ans. Coupling efficiency increases with both core diameter and Δ, while bending losses increase directly with core diameter and inversely with Δ.

Q. 15. *While coupled power and bending loss favor a high Δ, which Δ value, smaller or larger, improves fiber bandwidth?*

Ans. Smaller.

Q. 16. *What are the two basic types of single mode step-index fibers?*

Ans. Matched-clad and depressed-clad.

Q. 17. *Which fiber cladding, matched or depressed, consists of two regions?*

Ans. Depressed.

Q. 18. *In single mode operation, the value of the normalized frequency (V) should remain near the 2.405 level. If the value of V is less than 1, do single mode fibers carry a majority of the power in the core or cladding material?*

Ans. Cladding material.

Q. 19. *What happens to the fundamental mode as the operating wavelength becomes longer than the single mode cutoff wavelength?*

Ans. The fundamental mode becomes increasingly lossy.

Q. 20. *Give two reasons why is the value of the normalized frequency (V) is varied in single mode step-index fibers?*

Ans. To increase performance and reduce losses caused by bending and splicing.

Q. 21. *Give two reasons why optical fiber manufacturers depart from the traditional circular core and cladding, low-loss glass fiber design?*

Ans. To increase performance and reduce cost.

Q. 22. *What five characteristics do applications using plastic clad silica (PCS) and all-plastic fibers typically have?*

Ans. High NA, low bandwidth, tight bend radius, short length, and low cost.

Q. 23. *List the types of materials used in fabricating low-loss, long wavelength optical fibers.*

Ans. Heavy-metal fluorides, chalcogenide glasses, and crystalline materials.

Q. 24. *What are the two methods used by fiber manufacturers to fabricate multimode and single mode glass fibers?*

Ans. Vapor phase oxidation and direct-melt process.

Q. 25. *Which method, vapor phase oxidation or direct-melt process, transforms deposited material into a solid glass preform by heating the porous material without melting?*

Ans. Vapor phase oxidation.

Q. 26. *List three benefits that properly cabled optical fibers provide.*

Ans.
- Protect optical fibers from damage or breakage during installation and over the fiber's lifetime.
- Provide stable fiber transmission characteristics compared with uncabled fibers.
- Maintain the physical integrity of the optical fiber.

Q. 27. *In addition to a primary coating, manufacturers add a layer of buffer material for what reasons?*

Ans. To provide additional mechanical protection and preserve the fiber's inherent strength.

Q. 28. *List the three techniques used by manufacturers to buffer optical fibers.*

Ans. Tight-buffered, loose-tube, and gel-filled loose-tube.

Q. 29. *List seven properties cable jackets should have.*

Ans. Low smoke generation, low toxicity, low halogen content, flame retardance, fluid resistance, high abrasion resistance, and stable performance over temperature.

Q. 30. *List the three types of cable designs being considered by the Navy.*

Ans. Optical fiber cable component (OFCC), stranded, and ribbon cables designs.

Q. 31. *Describe an optical fiber cable component (OFCC).*

Ans. OFCCs are tight-buffer fiber surrounded by arimid yarn and a low-halogen outer jacket.

Q. 32. *Two layers of arimid yarn strength members encase the OFCC units. Why are these strength members stranded in opposing directions?*

Ans. To minimize microbending of the fibers.

Q. 33. *Why do cable manufacturers introduce a controlled twist to the stacked ribbons during the cabling process?*

Ans. To minimize fiber stress when the cable is bent.

Q. 34. *OFCC, stranded, and ribbon cables have different fiber capacities. What is the approximate number of fibers that each cable can accommodate in a 0.5-inch cable?*

Ans. OFCC (12 fibers), stranded (48 fibers), ribbon (204 fibers).

Q. 35. *Which fiber optic cable (OFCC, stranded, or ribbon) has the worst bend performance?*

Ans. Ribbon.

■■■

4 OPTICAL SPLICES, CONNECTORS AND COUPLERS

Learning Objectives

Upon completion of this chapter, you should be able to do the following:

- Describe a fiber optic splice, connector, and coupler and the types of connections they form in systems.
- List the types of extrinsic and intrinsic coupling losses.
- Understand the degree to which fiber alignment and fiber mismatch problems increase system loss.
- Identify the types of fiber optic mechanical and fusion splices. Outline the basic splicing techniques for each type of fiber optic splice.
- List the types of fiber optic connectors. Detail the procedure for installing a fiber optic connector on an optical fiber.
- Discuss the types of fiber optic passive couplers.

FIBER OPTIC CONNECTIONS

Chapter 1 states that a fiber optic data link performs three basic functions. First, the data link transmitter converts an electrical input signal to an optical signal. Then, the optical fiber transmits this optical signal. Finally, the data link receiver converts the optical signal back to an electrical signal identical to the original input. However, Chapter 1 does not describe how optical power transfers from one optical component to another.

This chapter describes how optical power is transferred from one fiber optic component to another. It describes how an optical source launches optical power into a fiber as well as how one optical fiber couples light into another fiber. In fiber optic system design, this launching or coupling of optical power from one component to the next is important.

Fiber optic connections permit the transfer of optical power from one component to another. Fiber optic connections also permit fiber optic systems to be more than just point-to-point data communication links. In fact, fiber optic data links are often of a more complex design than point-to-point data links.

A system connection may require either a fiber optic splice, connector or coupler. One type of system connection is a permanent connection made by splicing optical fibers together. A fiber optic **splice** makes a permanent joint between two fibers or two groups of fibers. There are two types of fiber optic splices—mechanical splices and fusion splices. Even though removal of some mechanical splices is possible, they are intended to be permanent. Another type of connection that allows for system reconfiguration is a fiber optic **connector.** Fiber optic connectors permit easy coupling and uncoupling of optical fibers. Fiber optic connectors sometimes resemble familiar electrical plugs and sockets. Systems may also divide or combine optical signals between fibers. Fiber optic **couplers** distribute or combine optical signals between fibers. Couplers can distribute an optical signal from a single fiber into several fibers. Couplers may also combine optical signals from several fibers into one fiber.

Fiber optic connection losses may affect system performance. **Poor fiber end preparation** and **poor fiber alignment** are the main causes of coupling loss. Another source of coupling loss is differences in optical properties between the connected fibers. If the connected fibers have different optical properties, such as different numerical apertures, core and cladding diameters, and refractive index profiles, then coupling losses may increase.

OPTICAL FIBER COUPLING LOSS

Ideally, optical signals coupled between fiber optic components are transmitted with no loss of light. However, there is always some type of imperfection present at fiber optic connections that causes some loss of light. It is the amount of optical power lost at fiber optic connections that is a concern of system designers.

The design of fiber optic systems depends on how much light is launched into an optical fiber from an optical source and how much light is coupled between fiber optic components, such as from one fiber to another. The amount of power launched from a source into a fiber depends on the optical properties of both the source and the fiber. The amount of optical power launched into an optical fiber depends on the radiance of the optical source. An optical source's **radiance** or brightness, is a measure of its optical power launching capability. Radiance is the amount of optical power emitted in a specific direction per unit time by a unit area of emitting surface. For most types of optical sources, only a fraction of the power emitted by the source is launched into the optical fiber.

The loss in optical power through a connection is defined similarly to that of signal attenuation through a fiber. Optical loss is also a log relationship. The loss in optical power through a connection is defined as:

$$\text{loss} = 10 \log_{10} \frac{P_i}{P_o}$$

For example, P_o is the power emitted from the source fiber in a fiber-to-fiber connection. P_i is the power accepted by the connected fiber. In any fiber optic connection, P_o and P_i are the optical power levels measured before and after the joint, respectively.

Fiber-to-fiber connection loss is affected by intrinsic and extrinsic coupling losses. **Intrinsic coupling losses** are caused by inherent fiber characteristics. **Extrinsic coupling losses** are caused by jointing techniques. Fiber-to-fiber connection loss is increased by the following sources of intrinsic and extrinsic coupling loss:

- Reflection losses
- Fiber separation
- Lateral misalignment
- Angular misalignment
- Core and cladding diameter mismatch
- Numerical aperture (NA) mismatch
- Refractive index profile difference
- Poor fiber end preparation

Intrinsic coupling losses are limited by reducing fiber mismatches between the connected fibers. This is done by procuring only fibers that meet stringent geometrical and optical specifications. Extrinsic coupling losses are limited by following proper connection procedures.

Some fiber optic components are modular devices that are designed to reduce coupling losses between components. Modular components can be easily inserted or removed from any system. For example, fiber optic transmitters and receivers are modular components. Fiber optic transmitters and receivers are devices that are generally manufactured with fiber pigtails or fiber optic connectors as shown in Fig. 4.1. A **fiber pigtail** is a short length of optical fiber (usually 1 meter or less) permanently fixed to the optical source or detector. Manufacturers supply transmitters and receivers with pigtails and connectors because fiber coupling to sources and detectors must be completed during fabrication. Reduced coupling loss results when source-to-fiber and fiber-to-detector coupling is done in a controlled manufacturing environment. Since optical sources and detectors are pigtailed or connectorized, launching

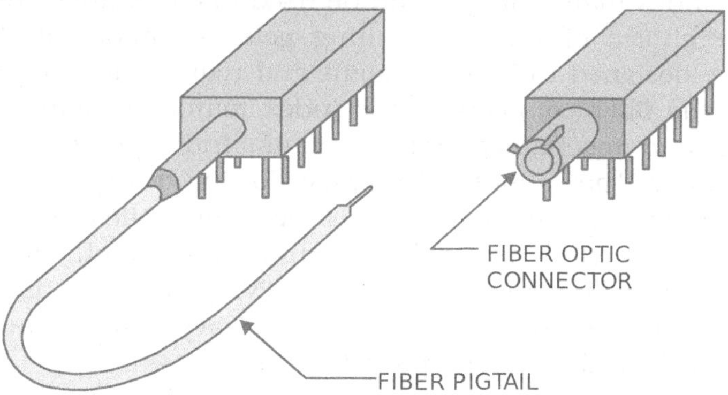

Fig. 4.1. *Pigtailed and connectorized fiber optic devices.*

optical power is reduced to coupling light from one fiber to another. In fact, most fiber optic connections can be considered fiber-to-fiber.

REFLECTION LOSSES

When optical fibers are connected, optical power may be reflected back into the source fiber. Light that is reflected back into the source fiber is lost. This reflection loss, called Fresnel reflection, occurs at every fiber interface. **Fresnel reflection** is caused by a step change in the refractive index that occurs at the fiber joint. In most cases, the step change in refractive index is caused by the ends of each fiber being separated by a small gap. This small gap is usually an air gap. In Fresnel reflection, a small portion of the incident light is reflected back into the source fiber at the fiber interface. The ratio (R), shown below, approximates the portion of incident light (light of normal incidence) that is reflected back into the source fiber.

$$R = \left(\frac{n_1 - n_0}{n_1 + n_0}\right)^2$$

R is the fraction of the incident light reflected at the fiber n_1 is the refractive index of the fiber core. n_0 is the refractive index of the medium between the two fibers.

Fresnel refraction occurs twice in a fiber-to-fiber connection.

A portion of the optical power is reflected when the light first exits the source fiber. Light is then reflected as the optical signal enters the receiving fiber. Fresnel reflection at each interface must be taken into account when calculating the total fiber-to-fiber coupling loss. Loss from Fresnel reflection may be significant. To reduce the amount of loss from Fresnel reflection, the air gap can be filled with an index matching gel. The refractive index of the index matching gel should match the refractive index of the fiber core. **Index matching gel** reduces the step change in the refractive index at the fiber interface, reducing Fresnel reflection.

In any system, index matching gels can be used to eliminate or reduce Fresnel reflection. The choice of index matching gels is important. Fiber-to-fiber connections are designed to be permanent and require no maintenance. Over the lifetime of the fiber connection, the index matching material must meet specific optical and mechanical requirements. Index matching gels should remain transparent. They should also resist flowing or dripping by remaining viscous. Some index matching gels darken over time while others settle or leak out of fiber connections. If these requirements are not met, then the fiber-to-fiber connection loss will increase over time. In Navy applications, this variation in connection loss over time is unacceptable. In Navy systems, index matching gels are only used in fiber optic splice interfaces.

FIBER ALIGNMENT

A main source of extrinsic coupling loss in fiber-to-fiber connections is poor fiber alignment. The three basic coupling errors that occur during fiber alignment are fiber separation (longitudinal misalignment), lateral misalignment, and angular misalignment. Most alignment errors are the result of mechanical imperfections introduced by fiber jointing techniques. However, alignment errors do result from installers not following proper connection procedures.

With **fiber separation,** a small gap remains between fiber-end faces after completing the fiber connection. Figure 4.2 illustrates this separation of the fiber-end faces.

Lateral or axial, misalignment occurs when the axes of the two fibers are offset in a perpendicular direction. Figure 4.3 shows this perpendicular offset of the axes of two connecting fibers.

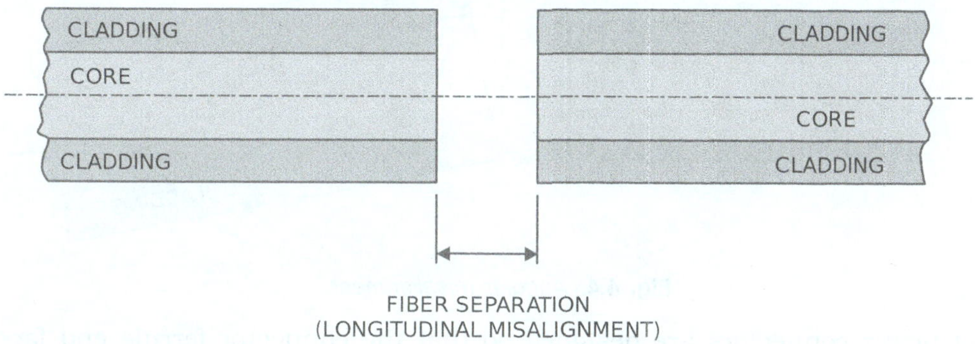

Fig. 4.2. *Fiber separation.*

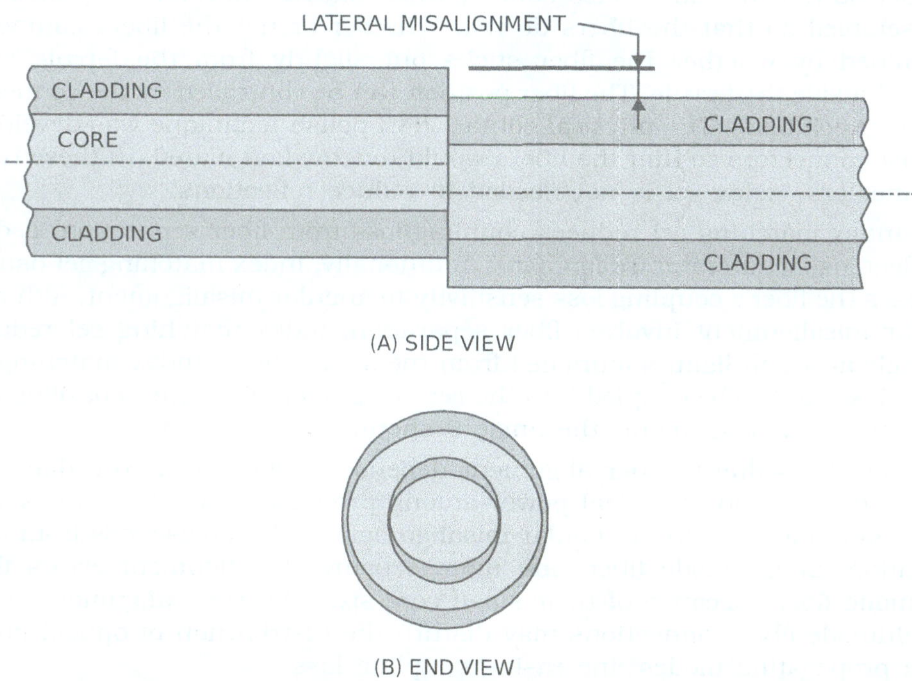

Fig. 4.3. *Lateral misalignment.*

Angular misalignment occurs when the axes of two connected fibers are no longer parallel. The axes of each fiber intersect at some angle (θ). Figure 4.4 illustrates the angular misalignment between the core axes.

Coupling loss caused by lateral and angular misalignment typically is greater than the loss caused by fiber separation. Loss, caused by fiber separation, is less critical because of the relative ease in limiting the distance of fiber separation. However, in some cases, fiber optic connectors prevent fibers from actual contact. These fiber optic connectors separate the fibers by a small gap. This gap eliminates damage to fiber-end faces during connection. For connectors with an air gap, the use of index matching gel reduces the coupling loss.

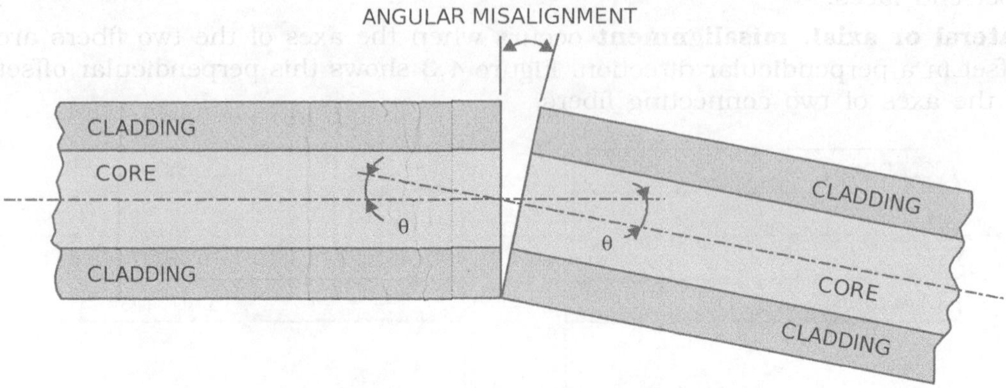

Fig. 4.4. *Angular misalignment.*

Most newer connectors are designed so that the connector ferrule end faces contact when the connector is mated. The connector can be assembled onto the fiber so that the fibers also contact when mated. However, they also can be assembled so that the fibers do not. Whether or not the fibers contact is determined by whether the fiber sticks out slightly from the ferrule or is recessed inside the ferrule. The fiber position can be controlled by the connector polishing technique. The physical contact (PC) polish technique was developed for most connectors so that the fibers would touch when mated. In these types of connectors, index gel is not needed to reduce reflections.

While index matching gel reduces coupling loss from fiber separation, it does not affect loss in lateral misalignment. Additionally, index matching gel usually increases the fiber's coupling loss sensitivity to angular misalignment. Although angular misalignment involves fiber separation, index matching gel reduces the angle at which light is launched from the source fiber. Index matching gel causes less light to be coupled into the receiving fiber. To reduce coupling loss from angular misalignment, the angle θ should be less than 1°.

Coupling losses due to fiber alignment depend on fiber type, core diameter, and the distribution of optical power among propagating modes. Fibers with large NAs reduce loss from angular misalignment and increase loss from fiber separation. Single mode fibers are more sensitive to alignment errors than multimode fibers because of their small core size. However, alignment errors in multimode fiber connections may disturb the distribution of optical power in the propagating modes, increasing coupling loss.

FIBER MISMATCHES

Fiber mismatches are a source of intrinsic coupling loss. As stated before, intrinsic coupling loss results from differences (mismatches) in the inherent fiber characteristics of the two connecting fibers. Fiber mismatches occur when manufacturers fail to maintain optical or structural (geometrical) tolerances during fiber fabrication.

Fiber mismatches are the result of inherent fiber characteristics and are independent of the fiber jointing techniques. Types of fiber mismatches include fiber geometry mismatches, NA mismatch and refractive index profile difference. Fiber geometry mismatches include core diameter, cladding diameter, core ellipticity, and core-cladding concentricity differences. Figure 4.5 illustrates each type of optical and geometrical fiber mismatch. Navy fiber specifications tightly specify these parameters to minimize coupling losses from fiber mismatches.

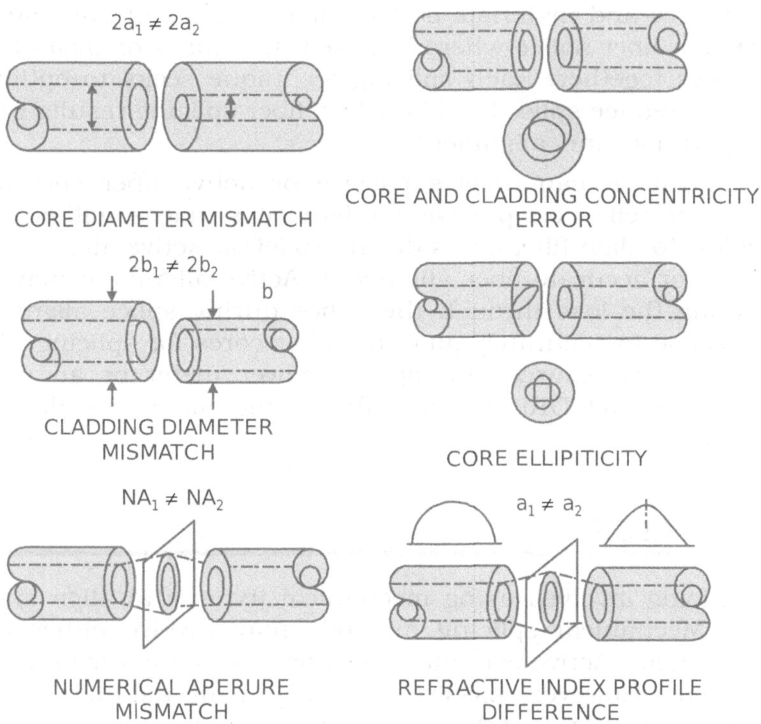

Fig. 4.5. *Types of optical and geometrical fiber mismatches that cause intrinsic coupling loss.*

Core diameter and NA mismatch have a greater effect on intrinsic coupling loss than the other types of fiber mismatches. In multimode fiber connections, the coupling loss resulting from core diameter mismatch, NA mismatch, and refractive index profile difference depends on the characteristics of the launching fiber. Coupling loss from core **diameter mismatch** results only if the launching fiber has a larger core radius (*a*) than the receiving fiber. Coupling loss from **NA mismatch** results only if the launching fiber has a higher NA than the receiving fiber. Coupling loss from **refractive index profile difference** results only if the launching fiber has a larger profile parameter (α) than the receiving fiber.

FIBER OPTIC SPLICES

A **fiber optic splice** is a permanent fiber joint whose purpose is to establish an optical connection between two individual optical fibers. System design may require that fiber connections have specific optical properties (low loss) that are met only by fiber-splicing. Fiber optic splices also permit repair of optical fibers damaged during installation, accident, or stress. System designers generally require fiber splicing whenever repeated connection or disconnection is unnecessary or unwanted.

Mechanical and fusion splicing are two broad categories that describe the techniques used for fiber splicing. A **mechanical splice** is a fiber splice where mechanical fixtures and materials perform fiber alignment and connection. A **fusion splice** is a fiber splice where localized heat fuses or melts the ends of two optical fibers together. Each splicing technique seeks to optimize splice performance and reduce splice loss. Low-loss fiber splicing results from proper fiber end preparation and alignment.

Fiber splice alignment can involve passive or active fiber core alignment. Passive alignment relies on precision reference surfaces, either grooves or cylindrical holes, to align fiber cores during splicing. Active alignment involves the use of light for accurate fiber alignment. Active alignment may consist of either monitoring the loss through the splice during splice alignment or by using a microscope to accurately align the fiber cores for splicing. To monitor loss either an optical source and optical power meter or an optical time domain reflectometer (OTDR) are used. Active alignment procedures produce low-loss fiber splices.

MECHANICAL SPLICES

Mechanical splicing involves using mechanical fixtures to align and connect optical fibers. Mechanical splicing methods may involve either passive or active core alignment. Active core alignment produces a lower loss splice than passive alignment. However, passive core alignment methods can produce mechanical splices with acceptable loss measurements even with single mode fibers.

In the strictest sense, a mechanical splice is a permanent connection made between two optical fibers. Mechanical splices hold the two optical fibers in alignment for an indefinite period of time without movement. The amount of splice loss is stable over time and unaffected by changes in environmental or mechanical conditions.

If high splice loss results from assembling some mechanical splices, the splice can be reopened and the fibers realigned. Realignment includes wiping the fiber or ferrule end with a soft wipe, reinserting the fiber or ferrule in a new arrangement, and adding new refractive index material. Once producing an acceptable mechanical splice, splice realignment should be unnecessary because most mechanical splices are environmentally and mechanically stable within their intended application.

The types of mechanical splices that exist for mechanical splicing include glass, plastic, metal, and ceramic tubes; and V-groove and rotary devices. Materials that assist mechanical splices in splicing fibers include transparent adhesives and index matching gels. **Transparent adhesives** are epoxy resins that seal mechanical splices and provide index matching between the connected fibers.

Glass or Ceramic Alignment Tube Splices

Mechanical splicing may involve the use of a glass or ceramic alignment tube, or capillary. The inner diameter of this glass or ceramic tube is only slightly larger than the outer diameter of the fiber. A transparent adhesive, injected into the tube, bonds the two fibers together. The adhesive also provides index matching between the optical fibers. Figure 4.6 illustrates fiber alignment using a glass or ceramic tube. This splicing technique relies on the inner diameter of the alignment tube. If the inner diameter is too large, splice loss will increase because of fiber misalignment. If the inner diameter is too small, it is impossible to insert the fiber into the tube.

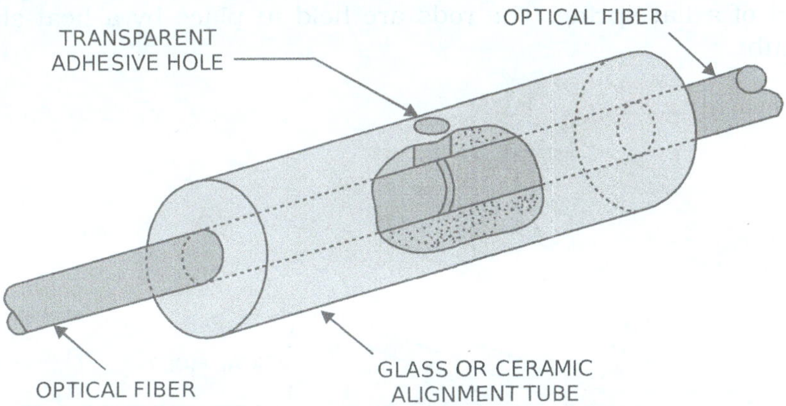

Fig. 4.6. *A glass or ceramic alignment tube for mechanical splicing.*

V-Grooved Splices

Mechanical splices may also use either a grooved substrate or positioning rods to form suitable V-grooves for mechanical splicing. The basic V-grooved device relies on an open grooved substrate to perform fiber alignment. When inserting the fibers into the grooved substrate, the V-groove aligns the cladding surface of each fiber end. A transparent adhesive makes the splice permanent by securing the fiber ends to the grooved substrate. Figure 4.7 illustrates this type of open V-grooved splice.

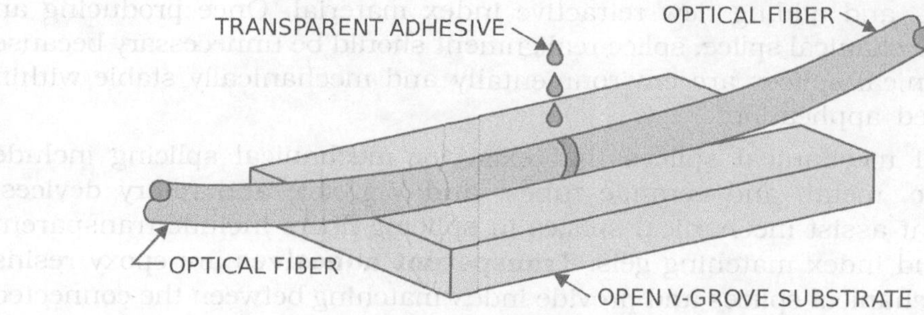

Fig. 4.7. *Open V-grooved splice.*

V-grooved splices may involve sandwiching the butted ends of two prepared fibers between a V-grooved substrate and a flat glass plate. Additional V-grooved devices use two or three positioning rods to form a suitable V-groove for splicing. The V-grooved device that uses two positioning rods is the spring V-grooved splice. This splice uses a groove formed by two rods positioned in a bracket to align the fiber ends. The diameter of the positioning rods permits the outer surface of each fiber end to extend above the groove formed by the rods. A flat spring presses the fiber ends into the groove maintaining fiber alignment. Transparent adhesive completes the assembly process by bonding the fiber ends and providing index matching. Figure 4.8 is an illustration of the spring V-grooved splice. A variation of this splice uses a third positioning rod instead of a flat spring. The rods are held in place by a heat-shrinkable band, or tube.

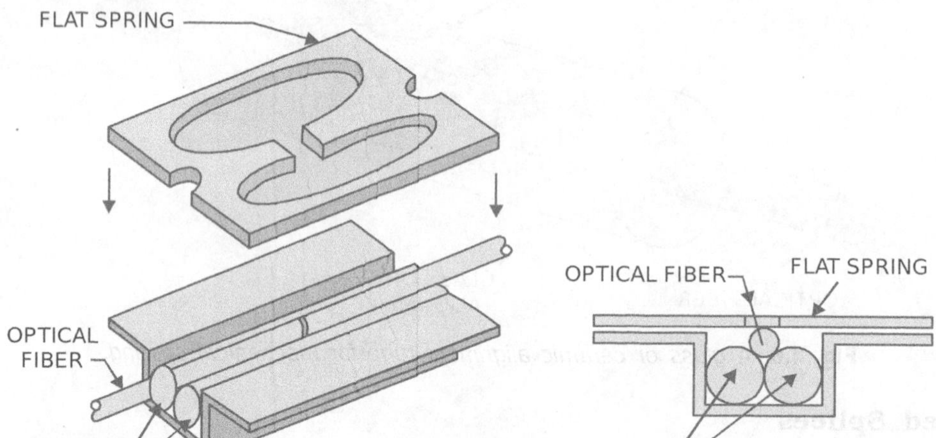

Fig. 4.8. *Spring V-grooved mechanical splice.*

Rotary Splices

In a rotary splice, the fibers are mounted into a glass ferrule and secured with adhesives. The splice begins as one long glass ferrule that is broken in half during the assembly process. A fiber is inserted into each half of the tube and epoxied in place using an ultraviolet cure epoxy. The endface of the tubes are then polished and placed together using the alignment sleeve. Figure 4.9 is an illustration of a rotary splice. The fiber ends retain their original orientation and have added mechanical stability since each fiber is mounted into a glass ferrule and alignment sleeve. The rotary splice may use index matching gel within the alignment sleeve to produce low-loss splices.

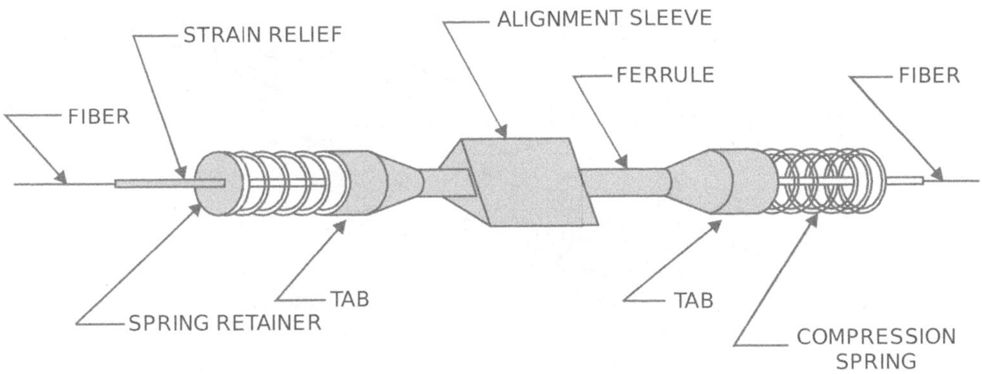

Fig. 4.9. *Rotary mechanical splice.*

In shipboard applications, the Navy recommends using the rotary splice. The rotary splice is a low-loss mechanical splice that provides stable environmental and mechanical performance in the Navy environment. Stable performance means that splice loss does not vary significantly with changes in temperature or other environmental or mechanical conditions. Completing a rotary splice also requires only a small amount of training, or expertise. This shorter training time is another reason why the Navy recommends using the rotary splice over other mechanical or fusion splicing techniques.

FUSION SPLICES

The process of fusion splicing involves using localized heat to melt or fuse the ends of two optical fibers together. The splicing process begins by preparing each fiber end for fusion. Fusion splicing requires that all protective coatings be removed from the ends of each fiber. The fiber is then cleaved using the score-and-break method. The quality of each fiber end is inspected using a microscope. In fusion splicing, splice loss is a direct function of the angles and quality of the two fiber-end faces.

The basic fusion splicing apparatus consists of two fixtures on which the fibers are mounted and two electrodes. Figure 4.10 shows a basic fusion-splicing apparatus. An inspection microscope assists in the placement of the prepared fiber ends into a fusion-splicing apparatus. The fibers are placed into the apparatus, aligned, and then fused together. Initially, fusion splicing used nichrome wire as the heating element to melt or fuse fibers together. New fusion-splicing techniques have replaced the nichrome wire with carbon dioxide (CO_2) lasers, **electric arcs** or gas flames to heat the fiber ends, causing them to fuse together. The small size of the fusion splice and the development of automated fusion-splicing machines have made **electric arc fusion** (arc fusion) one of the most popular splicing techniques in commercial applications.

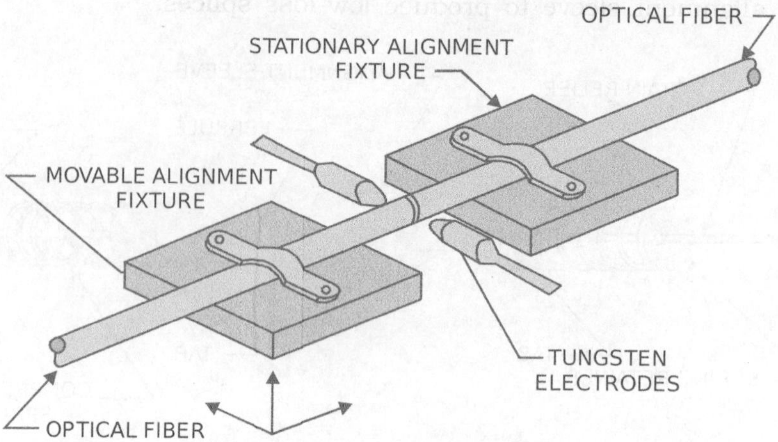

Fig. 4.10. *A basic fusion splicing apparatus.*

Arc fusion involves the discharge of electric current across a gap between two electrodes. By placing the fiber ends between the electrodes, the electric discharge melts or fuses the ends of each fiber. Figure 4.10 shows the placement of the fiber ends between tungsten electrodes during arc fusion. Initially, a small gap is present between the fiber ends. A short discharge of electric current is used to prepare the fiber ends for fusion. During this short discharge, known as **prefusion,** the fiber ends are cleaned and rounded to eliminate any surface defects that remain from fiber cleaving. Surface defects can cause core distortions or bubble formations during fiber fusion. A fusion splice results when the fiber ends are pressed together, actively aligned, and fused using a longer and stronger electric discharge. Automated fusion splicers typically use built-in local optical power launch/detection schemes for aligning the fibers. During fusion, the surface tension of molten glass tends to realign the fibers on their outside diameters, changing the initial alignment. When the fusion process is complete, a small core distortion may be present. Small core distortions have negligible effects on light propagating through multimode fibers. However, a small core distortion can significantly affect single mode fiber splice loss. The core distortion, and the splice loss, can be reduced by limiting the arc discharge and decreasing the gap distance between the two electrodes. This limits the region of molten glass. However, limiting the region of molten glass reduces the tensile strength of the splice.

Fusion splicing yields typically vary between 25 and 75 percent depending on the strength and loss requirements for the splice and other factors. Other factors affecting splice yields include the condition of the splicing machine, the experience of the splice personnel and environmental conditions. Since fusion splicing is inherently permanent, an unacceptable fusion splice requires breakage and refabrication of the splice.

In general, fusion splicing takes a longer time to complete than mechanical splicing. Also, yields are typically lower making the total time per successful splice much longer for fusion splicing. Both the yield and splice time are determined to a large degree by the expertise of the fusion splice operator. Fusion splice operators must be highly trained to consistently make low-loss reliable fusion splices. For these reasons the fusion splice is not recommended for use in Navy shipboard applications.

MULTIFIBER SPLICING

Normally, multifiber splices are only installed on ribbon type fiber optic cables. Multifiber splicing techniques can use arc fusion to restore connection, but most splicing techniques use mechanical splicing methods. The most common mechanical splice is the ribbon splice.

A ribbon splice uses an etched silicon chip, or grooved substrate, to splice the multiple fibers within a flat ribbon. The spacing between the etched grooves of the silicon chip is equal to the spacing between the fibers in the flat ribbon. Before placing each ribbon on the etched silicon chip, each fiber within the ribbon cable is cleaved. All of the fibers are placed into the grooves and held in place with a flat cover. Typically, an index matching gel is used to reduce the splice loss. Figure 4.11 shows the placement of the fiber ribbon on the etched silicon chip.

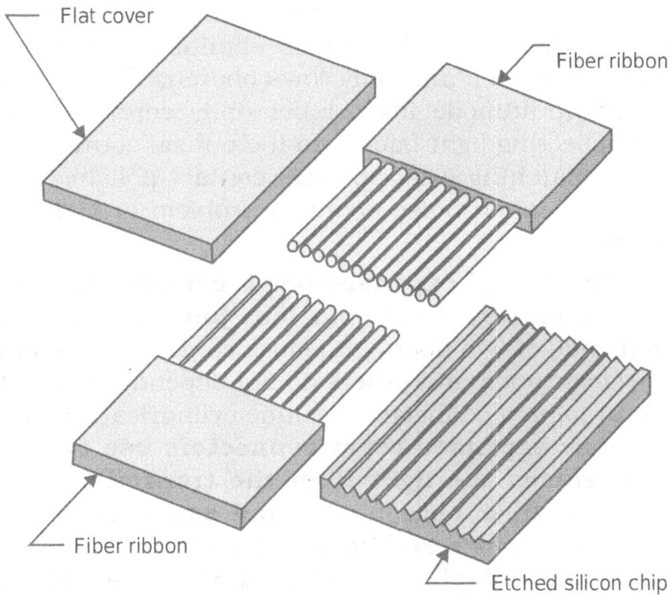

Fig. 4.11. *Ribbon splice on etched silicon chip.*

FIBER OPTIC CONNECTORS

A fiber optic connector is a demateable device that permits the coupling of optical power between two optical fibers or two groups of fibers. Designing a device that allows for repeated fiber coupling without significant loss of light is difficult. Fiber optic connectors must maintain fiber alignment and provide repeatable loss measurements during numerous connections. Fiber optic connectors should be easy to assemble (in a laboratory or field environment) and should be cost effective. They should also be reliable. Fiber optic connections using connectors should be insensitive to environmental conditions, such as temperature, dust, and moisture. Fiber optic connector designs attempt to optimize connector performance by meeting each of these conditions.

Fiber optic connector coupling loss results from the same loss mechanisms described earlier in this chapter. Coupling loss results from poor fiber alignment and end preparation (extrinsic losses), fiber mismatches (intrinsic loss), and Fresnel reflection. The total amount of insertion loss for fiber optic connectors should remain below 1 dB. Fiber alignment is the critical parameter in maintaining the total insertion loss below the required level. There is only a small amount of control over coupling loss resulting from fiber mismatches, because the loss results from inherent fiber properties. Index matching gels cannot be used to reduce Fresnel losses, since the index matching gels attract dust and dirt to the connection.

Fiber optic connectors can also reduce system performance by introducing modal and reflection noise. The cause of modal noise in fiber optic connectors is the interfering of the different wavefronts of different modes within the fiber at the connector interface. Modal noise is eliminated by using only single mode fiber with laser sources and only low-coherence sources such as light-emitting diodes with multimode fiber. Fiber optic connectors can introduce reflection noise by reflecting light back into the optical source. Reflection noise is reduced by index matching gels, physical contact polishes or antireflection coatings. Generally, reflection noise is only a problem in high data rate single mode systems using lasers.

Butt-jointed connectors and **expanded-beam connectors** are the two basic types of fiber optic connectors. Fiber optic **butt-jointed connectors** align and bring the prepared ends of two fibers into close contact. The end-faces of some butt-jointed connectors touch, but others do not depending upon the connector design. Types of butt-jointed connectors include cylindrical ferrule and biconical connectors. Fiber optic **expanded-beam connectors use two lenses to first expand and then refocus the light from the transmitting fiber into the receiving fiber. Single fiber butt-jointed and expanded beam connectors normally consist of two plugs and an adapter (coupling device). Fig. 4.12 shows how to configure each plug and adapter when making the connection between two optical fibers.**

Ferrule connectors use two cylindrical plugs (referred to as ferrules), an alignment sleeve, and sometimes axial springs to perform fiber alignment. Figure 4.13 provides an illustration of this basic ferrule connector design.

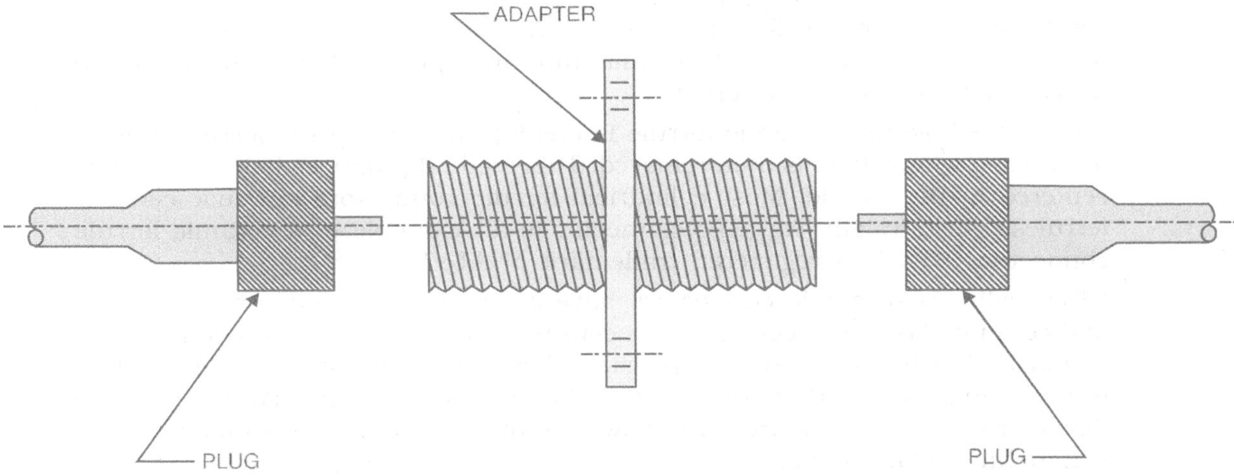

Fig. 4.12. *Plug-adapter-plug configuration.*

Precision holes drilled or molded through the center of each ferrule allow for fiber insertion and alignment. Precise fiber alignment depends on the accuracy of the central hole of each ferrule. When the fiber ends are inserted, an adhesive (normally an epoxy resin) bonds the fiber inside the ferrule. The fiber-end faces are polished until they are flush with the end of the ferrule to achieve a low-loss fiber connection. Fiber alignment occurs when the ferrules are inserted into the alignment sleeve. The inside diameter of the alignment sleeve aligns the ferrules, which in turn align the fibers. Ferrule connectors lock the ferrules in the alignment sleeve using a threaded outer shell or some other type of coupling mechanism.

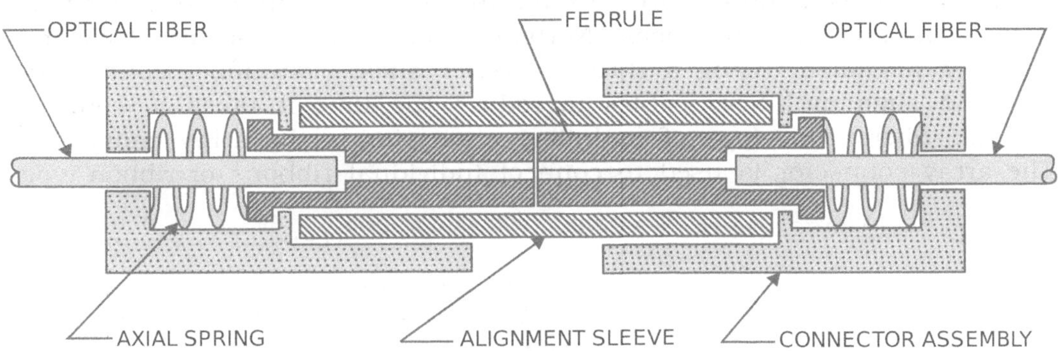

Fig. 4.13. *Basic ferrule connector design.*

As stated before, fiber alignment depends on an accurate hole through the center of the ferrule. Normally, ferrule connectors use ceramic or metal ferrules. The center hole is generally drilled in a metal ferrule. Drilling an accurate hole through the entire metal ferrule can be difficult. To improve fiber alignment, some metal ferrule connectors use precision watch-jeweled centering. In precision watch-jeweled centering, a watch jewel with a precision centered hole is placed in the tip of the ferrule. The central hole of the watch jewel centers the fiber with respect to the axis of the cylindrical ferrule. The watch jewel provides for better fiber alignment, because regulating the hole tolerance of the watch jewel is easier than maintaining a precise hole diameter when drilling through an entire ferrule.

The center hole in a ceramic ferrule is created by forming the ferrule around a precision wire, which is then removed. This method produces holes accurately centered in the ferrule. Most cylindrical ferrule connectors now use ceramic ferrules. The Straight Tip (ST®)connector is an example of a ceramic ferrule connector. (ST® is a registered trademark of AT&T.)

Other cylindrical ferrule connectors have a ferrule that contains both metal and ceramic. For these connectors a ceramic capillary is placed within the tip of a metal ferrule to provide for precision fiber alignment. The ceramic capillary is a ceramic tube with a small inner diameter that is just larger than the diameter of the fiber. Figure 4.14 shows the placement of the ceramic capillary within the metal ferrule.

Another type of butt-jointed connector is the biconical connector. Biconical connectors use two conical plugs, a double conical alignment sleeve, and axial springs to perform fiber alignment. Figure 4.15 is an illustration of this basic biconical connector design. Formation of the plugs and alignment sleeve involves transfer molding. Transfer molding uses silica-filled epoxy resin to mold the conical plug directly to the fiber or around a cast (precision wire). After connecting the conical plugs to the optical fibers, the fiber-end faces are polished before the plugs are inserted into the molded alignment sleeve. During fiber insertion, the inside surface of the double conical sleeve performs fiber alignment, while the axial springs push the fiber ends into close contact. If the alignment sleeve permits the fibers to actually become in contact, then the axial spring provides enough force to maintain fiber contact but prevent damage to the fiber-end faces. Normally, biconical connectors lock the fibers in alignment using a threaded outer shell. Multifiber connectors join and align multifiber cables to reduce the time it takes to connect multiple fibers. One type of multifiber connector is the array connector. The array connector is used to connect individual ribbons of ribbon-type

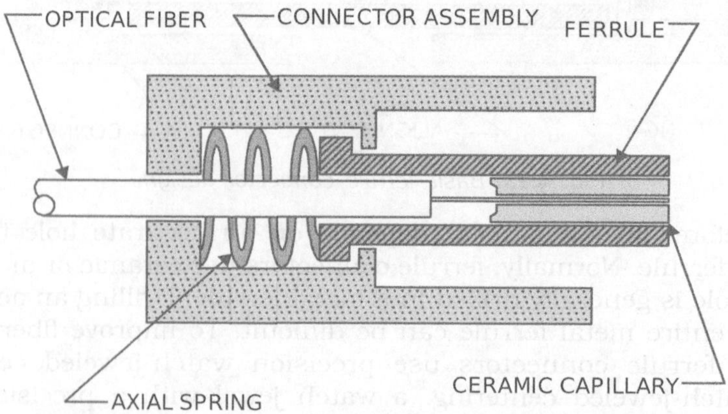

Fig. 4.14. *A ceramic capillary set within a metal ferrule.*

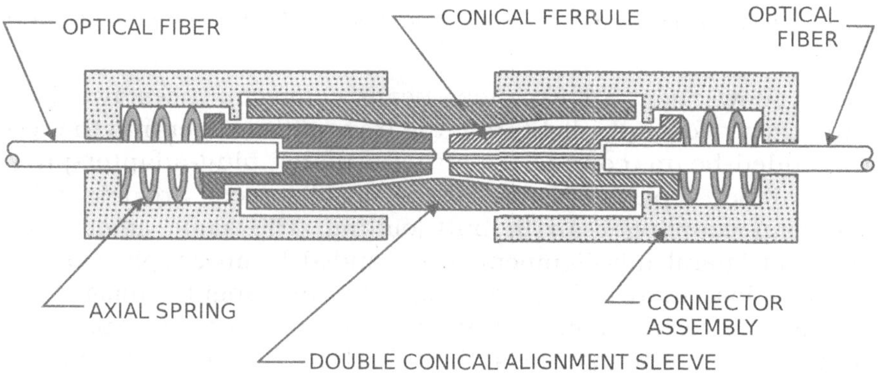

Fig. 4.15. *Biconical connector design.*

cables. The array connector is similar to the ribbon splice. In the array connector, the fibers of each ribbon are epoxied into grooves of a silicon chip so that the fiber ends protrude from the end of the chip. The chip and the protruding fibers are polished flat for connection. Each half of the connector is prepared separately before being butt-jointed. A spring clip and two grooved metal-backed plates are used to align and connect the stacked ribbons of the two ribbon cables. Array connectors may also use an alignment sleeve with V-grooved silicon chips and metal springs to align and connect stacked ribbons. Figure 4.16 shows the spring clip method of array connector alignment. The

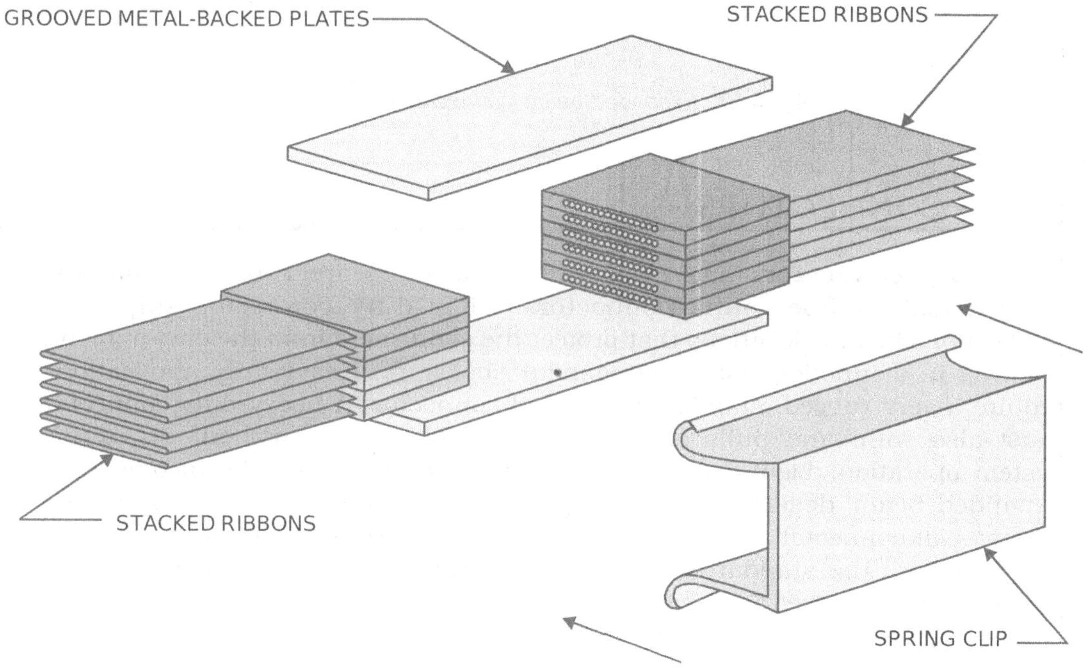

Fig. 4.16. *Spring clip method of ribbon connection.*

multifiber array connector is only one example of a multiple connector. Many types of multiple connectors exist that connect different types of multifiber cables.

Figure 4.17 shows how an expanded-beam connector uses two lenses to expand and then refocus the light from the transmitting fiber into the receiving fiber. Expanded-beam connectors are normally plug-adapter-plug type connections. Fiber separation and lateral misalignment are less critical in expanded-beam coupling than in butt-jointing. The same amount of fiber separation and lateral misalignment in expanded beam coupling produces a lower coupling loss than in butt-jointing. However, angular misalignment is more critical. The same amount of angular misalignment in expanded-beam coupling produces a higher loss than in butt-jointing. Expanded-beam connectors are also much harder to produce. Present applications for expanded-beam connectors include multifiber connections, edge connections for printed circuit boards, and other applications.

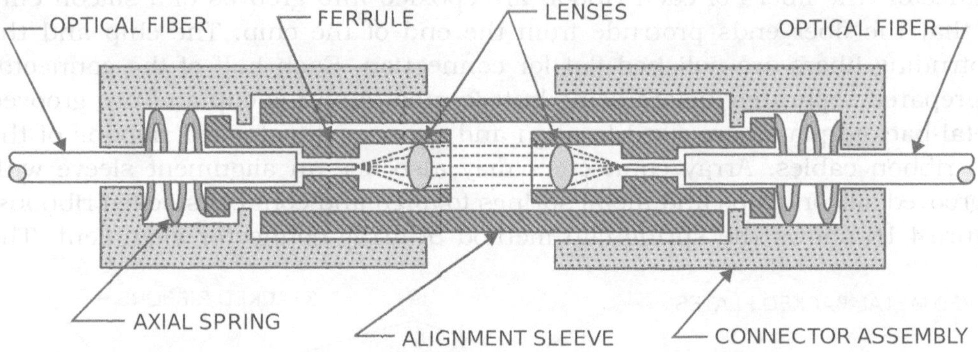

Fig. 4.17. *Expanded-beam connector operation.*

MILITARY CONNECTORS

Light-duty connectors and heavy-duty connectors are two ways that the Navy classifies fiber optic connectors. Light-duty connector shipboard applications include locations that protect the connectors from the environment, such as in a junction box or equipment enclosure. Heavy-duty applications require a very rugged, stand-alone, sealed connector. A heavy-duty connector must also withstand pulls and tugs on the fiber cable without disrupting system operation. Light-duty connectors can be of the ferrule, biconical, or expanded-beam designs. Ferrule-type ST® connectors are becoming the commercial connector of choice for local area network (LAN) and data transfer links and are the standard connector for Navy light duty applications. This connector is described in specification sheets 16, 17, and 18 of MIL-C-83522. Figure 4.18 shows the ST® type of light-duty connector.

Figure 4.19 shows one type of heavy-duty connector designed for use in harsh Navy environments. This connector is described by the military specification

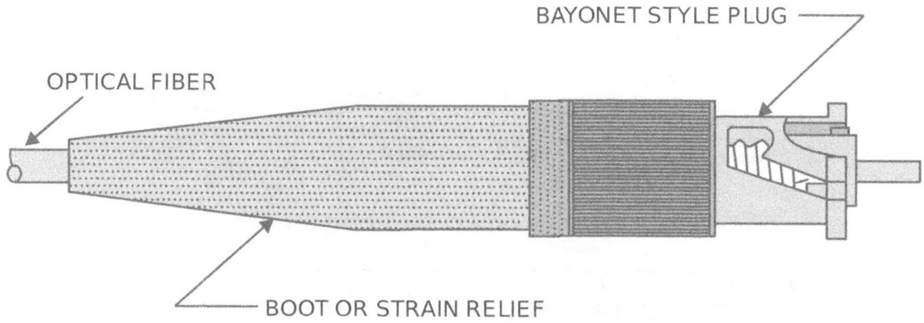

Fig. 4.18. *ST® light-duty connector.*

MIL-C-28876. This connector comes in various sizes capable of terminating 2, 4, 6, or 8 fibers. Each fiber termination, called a terminus, is of the cylindrical ferrule type. Two slightly different termini are used to form a connection; a pin terminus and a socket terminus. The pin terminus consists of a terminus body, which holds the terminus within the connector shell and a ceramic ferrule. The socket terminus consists of a terminus body, a ceramic ferrule, and an alignment sleeve, which attaches to the ceramic ferrule. Fiber alignment occurs when the pin terminus slides into the alignment sleeve of the socket terminus. The termini are held within an insert in the connector shell. When the connector halves are mated, the connector inserts align the mating termini, which then align the mating fibers. The connector shell and backshell protect the termini from the surrounding environment and provide strain relief for the multifiber cable.

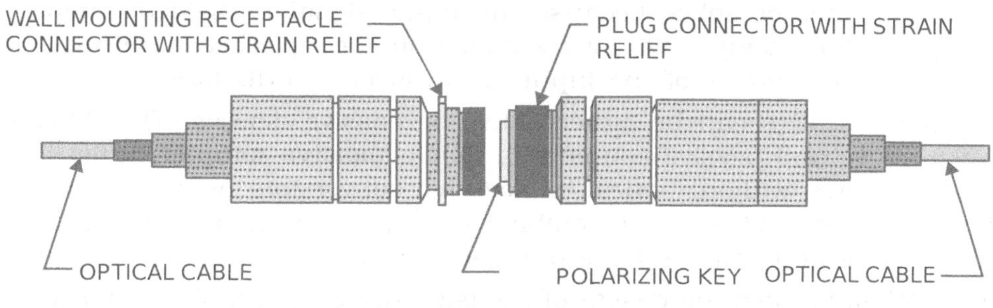

Fig. 4.19. *MIL-C-28876 heavy-duty connector.*

FIBER OPTIC COUPLERS

Some fiber optic data links require more than simple point-to-point connections. These data links may be of a much more complex design that requires multiport or other types of connections. Figure 4.20 shows some example system architectures that use more complex link designs. In many cases these types of systems require fiber optic components that can redistribute (combine or split) optical signals throughout the system.

One type of fiber optic component that allows for the redistribution of optical signals is a fiber optic coupler. A fiber optic coupler is a device that can

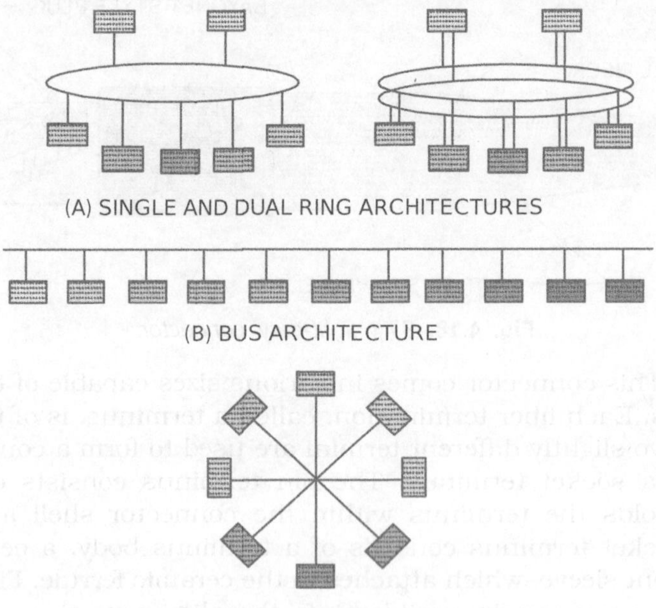

Fig. 4.20. *Examples of complex system architectures.*

distribute the optical signal (power) from one fiber among two or more fibers. A fiber optic coupler can also combine the optical signal from two or more fibers into a single fiber. Fiber optic couplers attenuate the signal much more than a connector or splice because the input signal is divided among the output ports. For example, with a 1 × 2 fiber optic coupler, each output is less than one-half the power of the input signal (over a 3 dB loss).

Fiber optic couplers can be either active or passive devices. The difference between active and passive couplers is that a **passive coupler** redistributes the optical signal without optical-to-electrical conversion. Active couplers are electronic devices that split or combine the signal electrically and use fiber optic detectors and sources for input and output.

Figure 4.21 illustrates the design of a basic fiber optic coupler. A basic fiber optic coupler has N input ports and M output ports. N and M typically range

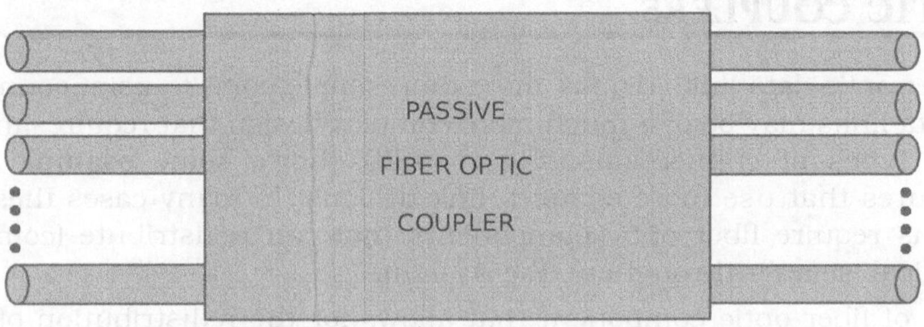

Fig. 4.21. *Basic passive fiber optic coupler design.*

from 1 to 64. The number of input ports and output ports vary depending on the intended application for the coupler. Types of fiber optic couplers include optical splitters, optical combiners, X-couplers, star couplers, and tree couplers.

An **optical splitter** is a passive device that splits the optical power carried by a single input fiber into two output fibers. Figure 4.22 illustrates the transfer of optical power in an optical splitter. The input optical power is normally split evenly between the two output fibers. This type of optical splitter is known as a **Y-coupler.** However, an optical splitter may distribute the optical power carried by input power in an uneven manner. An optical splitter may split most of the power from the input fiber to one of the output fibers. Only a small amount of the power is coupled into the secondary output fiber. This type of optical splitter is known as a T-coupler, or an optical tap.

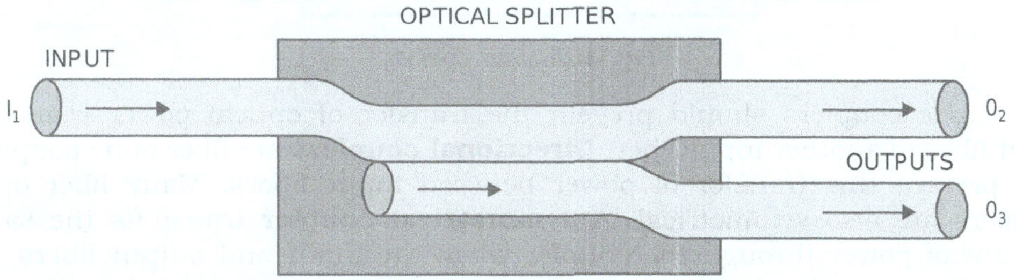

Fig. 4.22. *Optical splitter.*

An **optical combiner** is a passive device that combines the optical power carried by two input fibers into a single output fiber. Figure 4.23 illustrates the transfer of optical power in an optical combiner.

An **X-coupler** combines the functions of the optical splitter and combiner. The X-coupler combines and divides the optical power from the two input fibers between the two output fibers. Another name for the X-coupler is the 2 × 2 coupler.

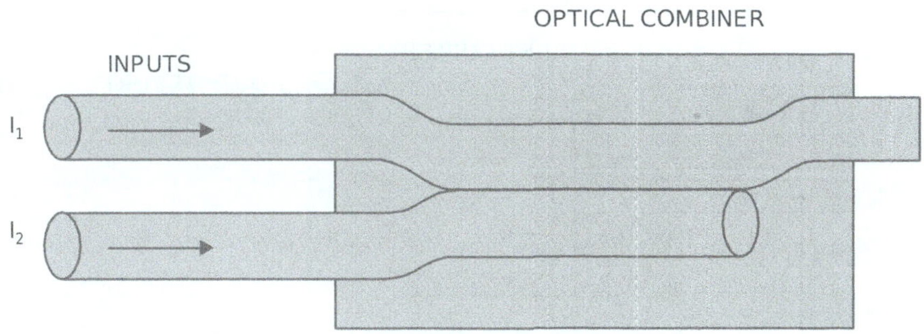

Fig. 4.23. *Optical combiner.*

Star and tree couplers are multiport couplers that have more than two input or two output ports. A **star coupler** is a passive device that distributes optical power from more than two input ports among several output ports. Figure 4.24 shows the multiple input and output ports of a star coupler. A **tree coupler** is a passive device that splits the optical power from one input fiber to more than two output fibers. A tree coupler may also be used to combine the optical power from more than two input fibers into a single output fiber. Figure 4.25 illustrates each type of tree coupler. Star and tree couplers distribute the input power uniformly among the output fibers.

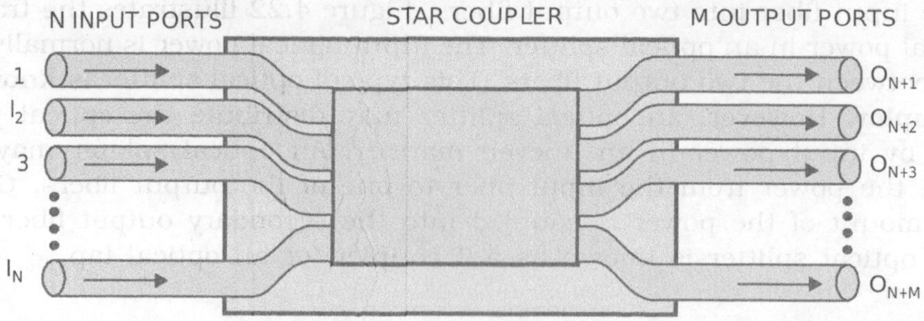

Fig. 4.24. *Star coupler.*

Fiber optic couplers should prevent the transfer of optical power from one input fiber to another input fiber. **Directional couplers** are fiber optic couplers that prevent this transfer of power between input fibers. Many fiber optic couplers are also symmetrical. A **symmetrical coupler** transmits the same amount of power through the coupler when the input and output fibers are reversed.

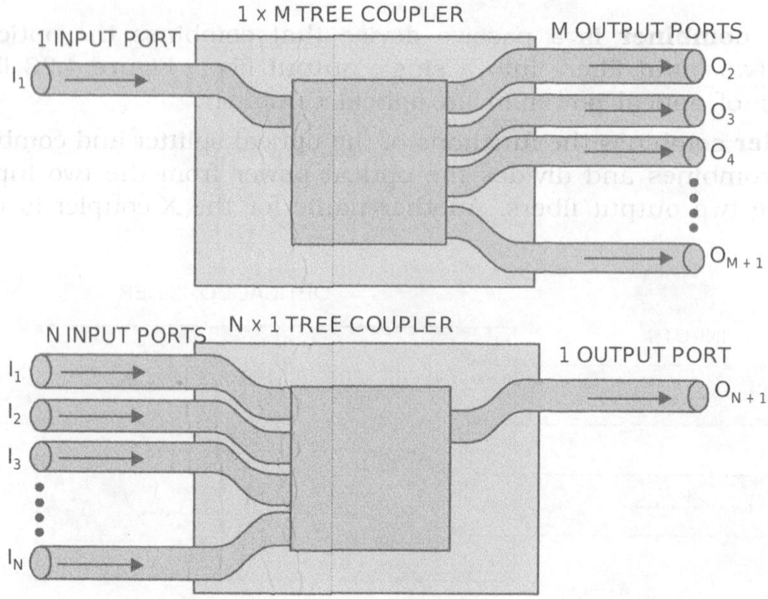

Fig. 4.25. *(1 × M) and (N × 1) tree coupler designs.*

Passive fiber optic coupler fabrication techniques can be complex and difficult to understand. Some fiber optic coupler fabrication involves beam splitting using microlenses or graded-refractive-index (GRIN) rods and beam splitters or optical mixers. These beamsplitter devices divide the optical beam into two or more separated beams. Fabrication of fiber optic couplers may also involve twisting, fusing, and tapering together two or more optical fibers. This type of fiber optic coupler is a fused biconical taper coupler. Fused biconical taper couplers use the radiative coupling of light from the input fiber to the output fibers in the tapered region to accomplish beam splitting. Figure 4.26 illustrates the fabrication process of a fused biconical taper coupler.

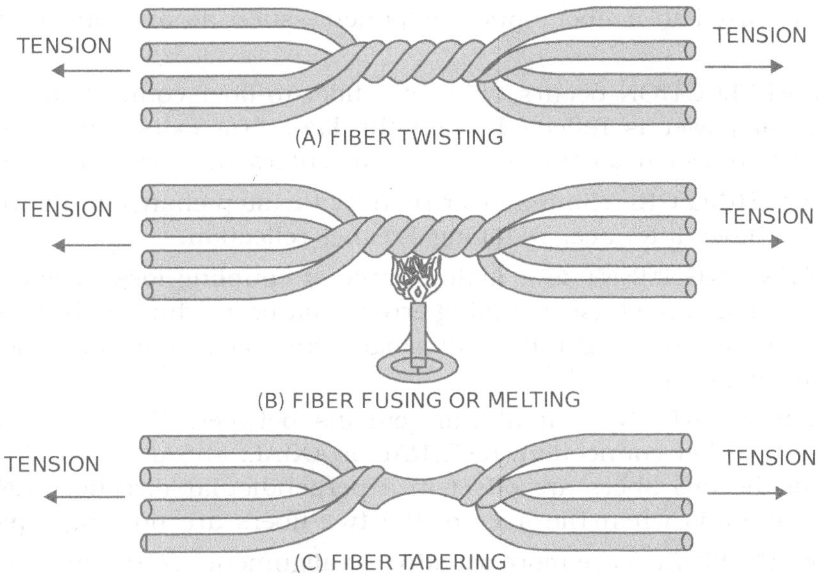

Fig. 4.26. *Fabrication of a fused biconical taper coupler (star coupler).*

SUMMARY

Now that you have completed this chapter, let's review some of the new terms, concepts, and ideas you have learned. You should have a thorough understanding of these principles before moving on to Chapter 5.

FIBER OPTIC CONNECTIONS transfer optical power from one component to another. Fiber optic connections also permit fiber optic systems to be more than just a point-to-point data link.

A FIBER OPTIC SPLICE is a permanent joint between two fibers or two groups of fibers.

FIBER OPTIC CONNECTORS permit easy coupling and uncoupling of optical fibers.

FIBER OPTIC COUPLERS distribute or combine optical signals between fibers.

POOR FIBER END PREPARATION and POOR FIBER ALIGNMENT are the main causes of coupling loss.

RADIANCE is the amount of optical power emitted by a unit area of emitting surface per unit time in a specified direction. An optical source's radiance, or brightness, is a measure of its optical power launching capability.

FIBER-TO-FIBER COUPLING LOSS is affected by intrinsic and extrinsic coupling losses. INTRINSIC COUPLING LOSSES are caused by inherent fiber characteristics. EXTRINSIC COUPLING LOSSES are caused by jointing techniques.

A FIBER PIGTAIL is a short length of optical fiber (usually 1 meter or less) permanently fixed to a fiber optic component, such as an optical source or detector.

FRESNEL REFLECTION occurs twice in a fiber-to-fiber connection. A portion of the optical power is reflected when the light first exits the source fiber. Light is then reflected as the optical signal enters the receiving fiber.

INDEX MATCHING GEL eliminates or reduces the step change in the refractive index at the fiber interface, reducing Fresnel reflection.

POOR FIBER ALIGNMENT is a main source of coupling loss in fiber-to-fiber connections. The three basic coupling errors that occur during fiber alignment are fiber separation (longitudinal misalignment), lateral misalignment, and angular misalignment.

In FIBER SEPARATION a small gap remains between fiber-end faces after completing the fiber connection. LATERAL or AXIAL, MISALIGNMENT is when the axes of the two fibers are offset in a perpendicular direction. ANGULAR MISALIGNMENT is when the axes of the two fibers are no longer parallel.

SINGLE MODE FIBERS are more sensitive to alignment errors than multimode fibers because of their small core diameters and low numerical apertures.

The MODE POWER DISTRIBUTION (MPD) is the distribution of radiant power among the various modes propagating along the optical fiber.

POLISHING the fiber ends removes most surface imperfections introduced by the fiber cleaving or cutting process. Fiber polishing involves a step-down approach. The first step is to give the surface of the fiber end a rough polish. The next step involves giving the surface of the fiber end a fine polish.

FIBER MISMATCHES are a source of intrinsic coupling loss. Types of fiber mismatches include fiber geometry mismatches, NA mismatch, and refractive index profile difference.

FIBER GEOMETRY MISMATCHES include core diameter, cladding diameter, core ellipticity and core-cladding concentricity differences.

CORE DIAMETER MISMATCH causes coupling loss only if the launching fiber has a larger core radius than the receiving fiber.

NA MISMATCH causes coupling loss only if the launching fiber has a higher NA than the receiving fiber.

A REFRACTIVE INDEX PROFILE DIFFERENCE causes coupling loss only if the launching fiber has a larger profile parameter than the receiving fiber.

MECHANICAL and FUSION SPLICING are two broad categories that describe the techniques used for fiber splicing. A mechanical splice is a fiber splice where mechanical fixtures perform fiber alignment and connection. A fusion splice is a fiber splice where localized heat fuses or melts the ends of two lengths of optical fiber together.

In MECHANICAL SPLICING, mechanical fixtures hold the two optical fibers in alignment for an indefinite period of time without movement. The amount of splice loss is stable over time and unaffected by changes in environmental or mechanical conditions.

ARC FUSION involves the discharge of electric current across a gap between two electrodes. By placing the fiber end between the electrodes, the electric discharge melts or fuses the ends of the fibers.

PREFUSION involves a short discharge of electric current across the gap between the electrodes. In prefusion the fiber ends are cleaned and rounded to eliminate any surface defects that remain from fiber cleaving.

A FIBER OPTIC CONNECTOR is a demateable device that permits the coupling of optical power between two optical fibers or two groups of fibers.

FIBER ALIGNMENT in a fiber optic connector is the critical parameter in maintaining total insertion loss below the required level.

FIBER OPTIC CONNECTORS can affect system performance by increasing modal and reflection noise.

MODAL NOISE is eliminated by using only single mode fiber with laser sources and only low-coherence sources such as light-emitting diodes with multimode fiber.

REFLECTION NOISE is reduced by index matching gels, physical contact polishes or antireflection coatings.

BUTT-JOINTED and EXPANDED BEAM CONNECTORS are two ways to classify fiber optic connectors. Butt-jointed connectors bring the prepared ends of two fibers into close contact. Expanded beam connectors use two lenses to first expand and then refocus the light from the transmitting fiber into the receiving fiber.

LIGHT-DUTY and HEAVY-DUTY CONNECTORS are two ways that the Navy classifies fiber optic connectors. Light-duty connector shipboard applications include locations that protect the connectors from the environment such as in a junction box. Heavy-duty applications require a very rugged, stand-alone, sealed connector.

A PASSIVE COUPLER redistributes an optical signal without optical to electrical conversion.

An OPTICAL SPLITTER is a passive device that splits the optical power carried by a single input fiber into two output fibers.

An OPTICAL COMBINER is a passive device that combines the optical power from two input fibers into a single output fiber.

A STAR COUPLER is a passive device that distributes optical power from more than two input ports among several output ports.

A TREE COUPLER is a passive device that splits the optical power from one input fiber to more than two output fibers. A tree coupler may also be used to combine the optical power from more than two input fibers into a single output fiber.

DIRECTIONAL COUPLERS are fiber optic couplers that prevent the transfer of optical power from one input fiber to another input fiber.

A SYMMETRICAL COUPLER transmits the same amount of power through the coupler when the input and output fibers are reversed.

REVIEW QUESTIONS AND ANSWERS

Q. 1. *Which fiber optic component (splice, connector, or coupler) makes a permanent connection in a distributed system?*

Ans. Splice.

Q. 2. *What are the main causes of coupling loss?*

Ans. Poor fiber end preparation and poor fiber alignment.

Q. 3. *Define the loss in optical power through a connection.*

Ans. $\text{loss} = 10 \log_{10} \frac{P_i}{P_o}$

Q. 4. *Fiber-to-fiber coupling loss is affected by intrinsic and extrinsic coupling losses. Can intrinsic coupling losses be limited by limiting fiber mismatches?*

Ans. Yes.

Q. 5. *In fiber-to-fiber connections, Fresnel reflection is one source of coupling losses. Light is reflected back into the source fiber and is lost. What causes Fresnel reflection?*

Ans. A step change in refractive index that occurs at fiber joints, caused by fiber separation.

Q. 6. *Reduction of Fresnel reflection is possible by reducing the step change in the refractive index at the fiber interface. What material reduces the step change in refractive index at a fiber interface?*

Ans. Index matching gel.

Q. 7. *List the three basic errors that occur during fiber alignment.*

Ans. Fiber separation (longitudinal misalignment), lateral misalignment, and angular misalignment.

Q. 8. *When the axes of two connected fibers are no longer in parallel, the two connected fibers are in what kind of misalignment?*

Ans. Angular misalignment.

Q. 9. *How does index matching gel affect the amount of coupling loss caused by (a) fiber separation, (b) lateral misalignment, and (c) angular misalignment?*

Ans. (*a*) Reduces coupling loss, (*b*) does not change coupling loss, and (*c*) increases coupling loss.

Q. 10. *Which are more sensitive to alignment errors, single mode or multimode fibers?*

Ans. Single mode.

Q. 11. *Quality fiber-end preparation is essential for proper system operation. What properties must an optical fiber-end face have to ensure proper fiber connection?*

Ans. Be flat, smooth, and perpendicular to the fiber axis.

Q. 12. *What is the basic fiber cleaving technique for preparing optical fibers for coupling?*

Ans. Score-and-break.

Q. 13. *Using a standard microscope to inspect a fiber-end face, you observe that all parts of the fiber-end face are in focus at the same time. Is the fiber-end face flat, concave, or convex?*

Ans. Flat.

Q. 14. *List six types of fiber mismatches.*

Ans. Core diameter mismatch, cladding diameter mismatch, core ellipticity, core and cladding concentricity differences, NA mismatch, and refractive index profile differences.

Q. 15. *Does coupling loss from refractive index profile difference result when the receiving fiber has a larger profile parameter (α) than the transmitting fiber?*

Ans. No.

Q. 16. *Define a fiber optic splice.*

Ans. A permanent fiber jint whose purpose is to establish an optical connection between two individual optical fibers.

Q. 17. *Fiber splicing is divided into two broad categories that describe the techniques used for fiber splicing. What are they?*

Ans. Mechanical and fusion splicing.

Q. 18. *Describe a transparent adhesive.*

Ans. An epoxy resin that seals mechanical splices and provides index matching between the connected fibers.

Q. 19. *The Navy recommends using the rotary splice for what two reasons?*

Ans. It is a low-loss mechanical splice that provides stable environmental and mechanical performance in the Navy environment, and it requires only a small amount of training.

Q. 20. *What fiber property directly affects splice loss in fusion splicing?*

Ans. The angles and quality of the two fiber-end faces.

Q. 21. *List two reasons why fusion splicing is one of the most popular splicing techniques in commercial applications.*

Ans. The small size of the fusion splice and the development of automated fusion-splicing machines.

Q. 22. *What is a short discharge of electric current that prepares the fiber ends for fusion called?*

Ans. Prefusion.

Q. 23. *Do small core distortions formed by arc fusion's self-alignment mechanism have more of an affect on light propagating through multimode or single mode fibers?*

Ans. Single mode fibers.

Q. 24. *What connection properties result in fiber optic connector coupling loss?*

Ans. Poor fiber alignment and end preparation, fiber mismatches, and Fresnel reflection.

Q. 25. *Which is the more critical parameter in maintaining total insertion loss below the required level, fiber alignment or fiber mismatch?*

Ans. Fiber alignment.

Q. 26. *Fiber optic connectors can reduce system performance by increasing what two types of noise?*

Ans. Modal and reflection.

Q. 27. *Which type of fiber optic connector (butt-jointed or expanded beam) brings the prepared ends of two optical fibers into close contact?*

Ans. Butt-jointed connectors.

Q. 28. *Is coupling loss from fiber separation and lateral misalignment more critical in expanded-beam or butt-jointed connectors?*

Ans. Butt-jointed connectors.

Q. 29. *Is coupling loss from angular misalignment more critical in expanded beam or butt-jointed connectors?*

Ans. Expanded beam connectors.

Q. 30. *The Navy classifies fiber optic connectors in what two ways?*

Ans. Light-duty and heavy-duty connectors.

Q. 31. *What is the difference between passive and active fiber optic couplers?*

Ans. Passive couplers redistribute optical signals without optical-to-electrical conversion.

Q. 32. *Which type of optical splitter (Y-coupler or T-coupler) splits only a small amount of power from the input fiber to one of the output fibers?*

Ans. T-coupler.

Q. 33. *Describe a directional coupler.*

Ans. A fiber optic coupler that prevents the transfer of power between input fibers.

5 FIBER OPTIC MEASUREMENT TECHNIQUES

Learning Objectives

Upon completion of this chapter, you should be able to do the following:

- Identify the prime reasons for conducting fiber optic manufacturing laboratory and field measurements.
- Describe the optical fiber and optical connection laboratory measurements performed by the Navy to evaluate fiber optic component and system performance.
- Describe the near-field and far-field optical power distribution of an optical fiber.
- Describe optical fiber launch conditions and modal effects that affect optical fiber and optical connection measurements.
- Understand the term optical time-domain reflectometry and the interpretation of an optical time-domain reflectometer (OTDR) trace.
- Describe the procedure for locating a fiber fault using an OTDR.

FIBER OPTIC MEASUREMENTS

Fiber optic data links operate reliably if fiber optic component manufacturers and end users perform the necessary laboratory and field measurements. Manufacturers must test how component designs, material properties, and fabrication techniques affect the performance of fiber optic components. These tests can be categorized as design tests or quality control tests. Design tests are conducted during the development of a component. Design tests characterize the component's performance (optical, mechanical and environmental) in the intended application. Once the component performance is characterized, the manufacturer generally only conducts quality control tests. Quality control tests verify that the parts produced are the same as the parts the design tests were conducted on. When manufacturers ship fiber optic components, they provide quality control data detailing the results of measurements performed during or after component fabrication.

End users (equipment manufacturers, shipbuilders, maintenance personnel, test personnel and so on) should measure some of these parameters upon receipt before installing the component into the fiber optic data link. These tests determine if the component has been damaged in the shipping process.

In addition, end users should measure some component parameters after installing or repairing fiber optic components in the field. The values obtained can be compared to the system installation specifications. These measurements determine if the installation or repair process has degraded component performance and will affect data link operation.

Whenever a measurement is made, it should be made using a standard measurement procedure. For most fiber optic measurements, these standard procedures are documented by the Electronics Industries Association/ Telecommunications Industries Association (EIA/TIA). Each component measurement procedure is assigned a unique number given by EIA/TIA-455-X. The X is a sequential number assigned to that particular component test procedure. System level test procedures are assigned unique numbers given by EIA/TIA-526-X. Again the X is a sequential number assigned to that particular system test procedure.

LABORATORY MEASUREMENTS

Providing a complete description of every laboratory measurement performed by manufacturers and end users is impossible. This chapter only provides descriptions of optical fiber and optical connection measurements that are important to system operation. The list of optical fiber and optical connection laboratory measurements described in this chapter includes the following:

- Attenuation
- Cutoff wavelength (single mode)
- Bandwidth (multimode)
- Chromatic dispersion
- Fiber geometry
- Core diameter
- Numerical aperture (multimode)
- Mode field diameter (single mode)
- Insertion loss
- Return loss and reflectance.

End users routinely perform optical fiber measurements to measure fiber power loss and fiber information capacity. End users may also perform optical fiber measurements to measure fiber geometrical properties. Optical fiber power loss measurements include attenuation and cutoff wavelength. Optical fiber information capacity measurements include chromatic dispersion and bandwidth. Fiber geometrical measurements include cladding diameter, core diameter, numerical aperture and mode field diameter. Optical connection measurements performed by end-users in the laboratory include insertion loss and reflectance or return loss.

Attenuation

Attenuation is the loss of optical power as light travels along the fiber. It is a result of absorption, scattering, bending, and other loss mechanisms as described in Chapter 3. Each loss mechanism contributes to the total amount of fiber attenuation.

End users measure the total attenuation of a fiber at the operating wavelength (λ). The **total attenuation (A)** between an arbitrary point X and point Y located on the fiber is

$$A = 10 \log \frac{P_x}{P_y} \text{ dB}$$

P_x is the power output at point X. P_y is the power output at point Y. Point X is assumed to be closer to the optical source than point Y. The total amount of attenuation will vary with changes in wavelength λ.

The **attenuation coefficient (α)** or attenuation rate, is

$$\alpha = \frac{A}{L} \text{ dB/km}$$

L is the distance between points X and Y. α is a positive number because P_x is always larger than P_y. The attenuation coefficient will also vary with changes in λ.

CUTBACK METHOD

In laboratory situations, end users perform the cutback method for measuring the total attenuation of an optical fiber. The cutback method involves comparing the optical power transmitted through a long piece of test fiber to the power present at the beginning of the fiber.

The cutback method for measuring multimode fiber attenuation is EIA/TIA-455-46. The cutback method for measuring single mode fiber attenuation is EIA/TIA-455-78. The basic measurement process is the same for both of these procedures. The test method requires that the test fiber of known length (L) be cut back to an approximate 2 m length. This cutback causes the destruction of 2 m of fiber. This method requires access to both fiber ends. Each fiber end should be properly prepared to make measurements. EIA/TIA-455-57 describes how to properly prepare fiber ends for measurement purposes.

Figure 5.1 illustrates the cutback method for measuring fiber attenuation. The cutback method begins by measuring, with an optical power meter, the output power P1 of the test fiber of known length (L) (Fig. 5.1, view A). Without disturbing the input light conditions, the test fiber is cut back to an approximate 2 m length. The output power P_2 of the shortened test fiber is then measured (Fig. 5.1, view B). The fiber attenuation A_T and the attenuation coefficient α are then calculated.

LAUNCH CONDITIONS. Measurement personnel must pay attention to how optical power is launched into the fiber when measuring fiber attenuation. Different distributions of launch power (launch conditions) can result in different attenuation measurements. This is more of a problem with multimode fiber than single mode fiber. For single mode fiber, optical power must be launched only into the fundamental mode. This is accomplished using a mode filter on the fiber. For multimode fiber, the distribution of power among the modes of the fiber must be controlled. This is accomplished by controlling the launch spot size and angular distribution.

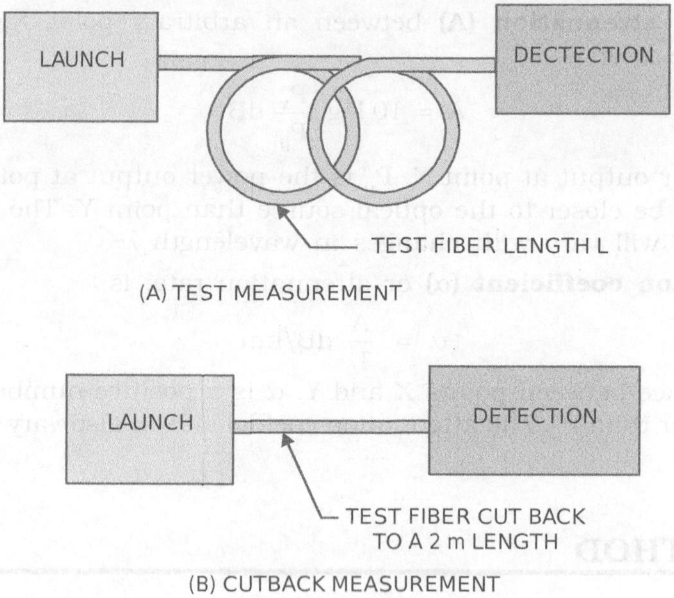

Fig. 5.1. *Cutback method for measuring fiber attenuation: A. Test measurement; B. Cutback measurement.*

The **launch spot size** is the area of the fiber face illuminated by the light beam from the optical source. The diameter of the spot depends on the size of the optical source and the properties of the optical elements (lenses, and so on) between the source and the fiber end face. The **angular distribution** is the angular extent of the light beam from the optical source incident on the fiber end face. The launch angular distribution also depends on the size of the optical source and the properties of the optical elements between the optical source and the fiber end face.

Multimode optical fiber launch conditions are typically characterized as being underfilled or overfilled. An underfilled launch concentrates most of the optical power in the center of the fiber. An **underfilled** launch results when the launch spot size and angular distribution are smaller than that of the fiber core. Underfilling the fiber excites mainly low-order modes.

Overfilling the fiber excites both low-order and high-order modes.

Figure 5.2 illustrates an overfilled launch condition.

An **overfilled** launch condition occurs when the launch spot size and angular distribution are larger than that of the fiber core. Incident light that falls outside the fiber core is lost. In addition, light that is incident at angles greater than the angle of acceptance of the fiber core is lost.

In attenuation measurements, cladding-mode strippers and mode filters eliminate the effects that high-order modes have on attenuation results. A **cladding-mode stripper** is a device that removes any cladding mode power from the fiber. Most cladding-mode strippers consist of a material with a

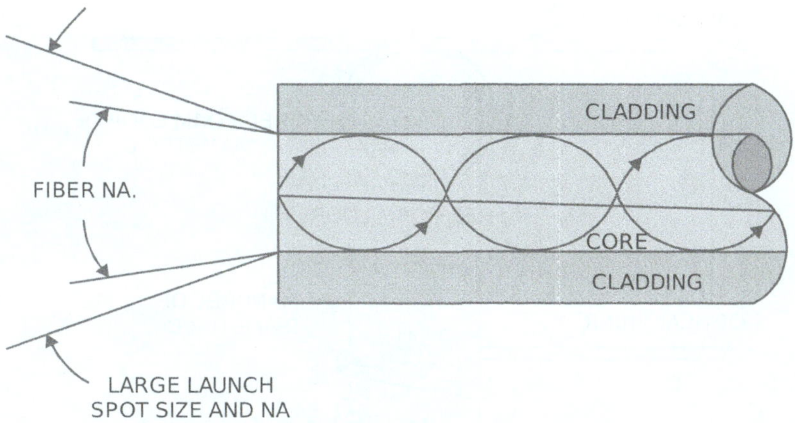

Fig. 5.2. *Overfilled launch condition.*

refractive index greater than that of the fiber cladding. For most fibers, the fiber coating acts as an excellent cladding-mode stripper.

A **mode filter** is a device that attenuates specific modes propagating in the core of an optical fiber. Mode filters generally involve wrapping the test fiber around a mandrel. For multimode, tight bends tend to remove high-order modes from the fiber. This type of mode filter is known as a **mandrel wrap mode filter.** For multimode fibers, mode filters remove high-order propagating modes and are individually tailored and adjusted for a specific fiber type.

For single mode fibers, a mode filter is used to eliminate the second-order mode from propagating along the fiber. The propagation of the second-order mode will affect attenuation measurements. Fiber attenuation caused by the second-order mode depends on the operating wavelength, the fiber bend radius and length.

The two most common types of mode filters are free-form loops and mandrel wraps. Figure 5.3 illustrates the free-form loop and mandrel-wrap types of mode filters. Mandrel wraps for multimode fibers consist of several wraps (approximately 4 or 5) around a mandrel. A 20 mm diameter mandrel is typically used for 62.5 μm fiber. Mandrel wraps for single mode fibers consist of a single wrap around a 30 mm diameter mandrel. Another common mode filter for single mode fibers is a 30 mm diameter circular free-form loop. Additional information on multimode and single mode filters (and launch conditions) is available in EIA/TIA-455-50 and EIA/TIA-455-77, respectively.

Launch conditions significantly affect the results of multimode fiber attenuation measurements. If the fiber is underfilled, high-order-mode power loss has minimal effect on the measurement results. If too much power is launched into high-order modes, the high-order-mode power loss will dominate the attenuation results. Generally, fiber attenuation measurements are performed using an underfilled launch condition. Power in high-order modes is eliminated by either controlling the input spot size and angular distribution or using mode filters to remove high-order mode power.

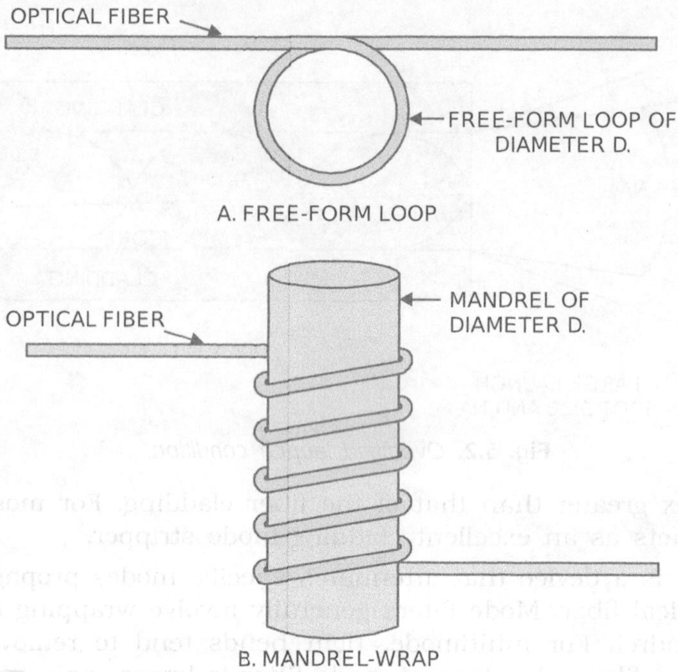

Fig. 5.3. *Types of mode filters: A. Free-form loop; B. Mandrel-wrap.*

Cutoff Wavelength

The wavelength at which a mode ceases to propagate is called the **cutoff wavelength** for that mode. However, an optical fiber is always able to propagate at least one mode, the fundamental mode. The fundamental mode can never be cutoff. The cutoff wavelength of a single mode fiber is the wavelength above which the fiber propagates only the fundamental mode.

Determining the cutoff wavelength of a single mode fiber involves finding the wavelength above which the power transmitted through the fiber decreased abruptly. This power decrease occurs when the second-order mode propagating in the fiber is cut off. The cutoff wavelength of single mode fibers depends on the fiber length and bend conditions. The effects of length and bending are different on different fibers depending on whether they are matched-clad or depressed-clad in design. The cutoff wavelength of matched-clad fibers is more sensitive to bends than the cuttoff wavelength of depressed-clad fibers. The cutoff wavelength of depressed-clad fibers is more sensitive to length than the cutoff wavelength of matched-clad fibers.

Cutoff wavelength may be measured on uncabled or cabled single mode fibers. A slightly different procedure is used in each case, but the basic measurement process is the same. The test method for uncabled single mode fiber cutoff wavelength is EIA/TIA-455-80. The test method for cabled single mode fiber cutoff wavelength is EIA/TIA-455-170. The fiber cutoff wavelength (λ_{cf}) measured under EIA/TIA-455-80 will generally be higher than the cable cutoff wavelength (λcc) measured under EIA/TIA-455-170. The difference is due to the fiber bends introduced during the cable manufacturing process.

Each test method describes the test equipment (input optics, mode filters, and cladding-mode strippers) necessary for the test. Cutoff wavelength measurements require an overfilled launch over the full range of test wavelengths. Since the procedures for measuring the cutoff wavelength of uncabled and cabled single mode fibers are essentially the same, only the test method for measuring the cutoff wavelength of uncabled fiber is discussed.

Measuring the cutoff wavelength involves comparing the transmitted power from the test fiber with that of a reference fiber at different wavelengths. The reference fiber can be the same piece of single mode fiber with small bends introduced or a piece of multimode fiber. If the same fiber with small bends is used as the reference fiber, the technique is called the bend-reference technique. If a piece of multimode fiber is used as the reference fiber, the technique is called the **multimode-reference technique.**

For both techniques, the test fiber is loosely supported in a single-turn with a constant radius of 140 mm. Figure 5.4 shows this single-turn configuration. The transmitted signal power $P_s(\lambda)$ is then recorded while scanning the wavelength range in increments of 10 nm or less. The launch and detection conditions are not changed while scanning over the range of wavelengths. The wavelength range scanned encompasses the expected cutoff wavelength.

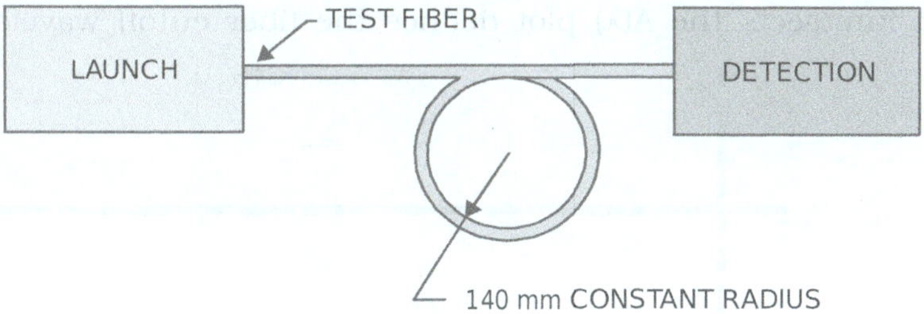

Fig. 5.4. *Single-turn configuration for the test fiber.*

The reference power measurement is then made. For the bend-reference technique, the launch and detection conditions are not changed, but an additional bend is added to the test fiber. The test fiber is bent to a radius of 30 mm or less to suppress the second-order mode at all the scanned wavelengths. For the multimode-reference technique, the single mode fiber is replaced with a 2 m length of multimode fiber. The transmitted signal power $P_r(\lambda)$ is recorded while scanning the same wavelength range in the same increments of 10 nm or less. The attenuation $A(\lambda)$ at each wavelength is calculated as follows:

$$A(\lambda) = 10 \log \frac{P_s(\lambda)}{P_r(\lambda)} \text{ dB}$$

Figure 5.5 shows an example attenuation plot generated using the bend-reference technique. The longest wavelength at which $A(\lambda)$ is equal to 0.1 dB is the fiber cutoff wavelength (λ_{cf}). λ_{cf} is marked in Fig. 5.5.

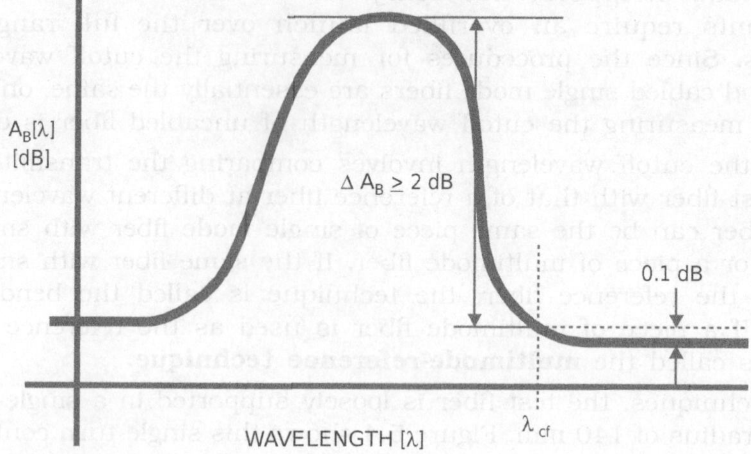

Fig. 5.5. *Fiber cutoff wavelength determined by the bend-reference technique.*

Figure 5.6 shows an attenuation plot generated using the multimode-reference technique. A straight line is fitted to the long-wavelength portion of A (λ). This straight line is then displaced upward by 0.1 dB. The point at which the straight line intersects the A(λ) plot defines the fiber cutoff wavelength (λ_{cf}).

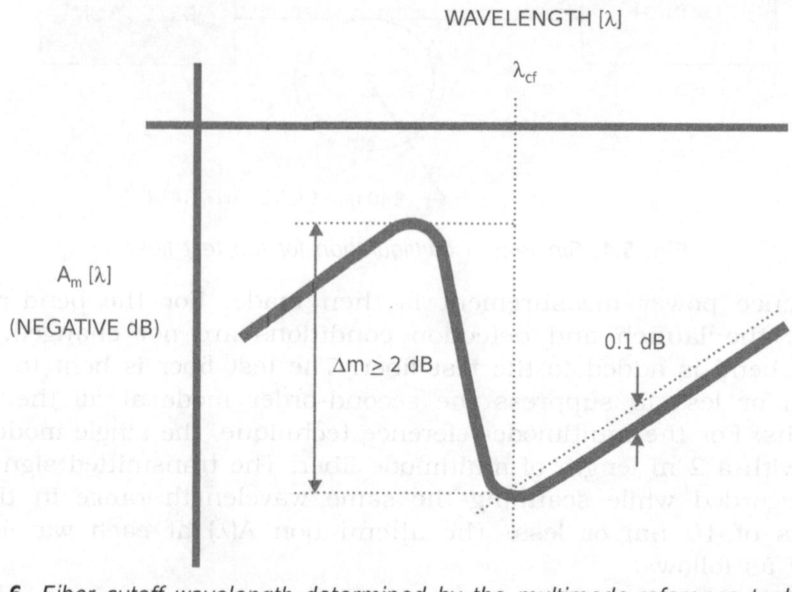

Fig. 5.6. *Fiber cutoff wavelength determined by the multimode-reference technique.*

Bandwidth

Dispersion reduces the bandwidth, or information-carrying capacity, of an optical fiber. **Dispersion** causes the spreading of the light pulse as it travels along the fiber (see Fig. 2.20). Fiber dispersion mechanisms include intramodal (chromatic) dispersion and intermodal (modal) dispersion. Multimode fiber bandwidth is a measure of the intermodal dispersion of the multimode fiber.

Intermodal dispersion is maximum when all fiber modes are excited. The source used for intermodal dispersion measurements must overfill the fiber. The optical source must also have a narrow spectral width to reduce the effects of chromatic dispersion in the measurement.

There are two basic techniques for measuring the modal bandwidth of an optical fiber. The first technique characterizes dispersion by measuring the **impulse response *h*(*t*)** of the fiber in the time domain. The second technique characterizes modal dispersion by measuring the **baseband frequency response H (*f*)** of the fiber in the frequency domain. H(*f*) is the **power transfer function** of the fiber at the baseband frequency (*f*). H(*f*) is also the Fourier transform of the power impulse response *h*(*t*). Only the Frequency response method is described here.

The test method for measuring the bandwidth of multimode fibers in the frequency domain is EIA/TIA-455-30. Signals of varying frequencies (*f*) are launched into the test fiber and the power exiting the fiber at the launched fundamental frequency measured. This optical output power is denoted as $P_{out}(f)$. The test fiber is then cut back or replaced with a short length of fiber of the same type. Signals of the same frequency are launched into the cut-back fiber and the power exiting the cut-back fiber at the launched fundamental frequency measured. The optical power exiting the cutback or replacement fiber is denoted as $P_{in}(f)$. The magnitude of the optical fiber frequency response is

$$H(f) = \log_{10}\left[\frac{P_{out}(f)}{P_{in}(f)}\right]$$

The fiber bandwidth is defined as the lowest frequency at which the magnitude of the fiber frequency response has decreased to one-half its zero-frequency value. This is the –3 decibel (dB) optical power frequency ($f_{3\ dB}$). This frequency is referred to as the fiber bandwidth.

Bandwidth is normally given in units of megahertz-kilometers (MHz-km). Converting the –3 dB fiber bandwidth to a unit length assists in the analysis and comparison of optical fiber performance. For long lengths of fiber (>1 km), the method for normalization is to multiply the length times the measured bandwidth.

Chromatic Dispersion

Chromatic, or intramodal, dispersion occurs in both single mode and multimode optical fibers. Chromatic dispersion occurs because different colors of light travel through the fiber at different speeds. Since the different colors of light have different velocities, some colors arrive at the fiber end before others. This delay difference is called the differential group delay $\tau(\lambda)$ per unit length. This differential group delay leads to pulse broadening.

Chromatic dispersion is measured using EIA/TIA-455-168 in the time domain. Chromatic dispersion is also measured in the frequency domain using EIA/TIA-455-169 and EIA/TIA-455-175. These methods measure the composite optical fiber material and waveguide dispersion. To understand the contribution that material and waveguide dispersive mechanisms have on multimode and single mode fiber dispersion, refer to Chapter 2. In this chapter we limit the discussion on chromatic dispersion to the time domain method described in EIA/TIA-455-168.

The chromatic dispersion of multimode graded-index and single mode fiber is obtained by measuring fiber group delays in the time domain. These measurements are made using multiwavelength sources or multiple sources of different wavelengths. A multiwavelength source could be a wavelength-selectable laser.

The pulse delay for both a long test sample fiber and a short reference fiber are measured over a range of wavelengths. The pulse delay for the reference fiber as a function of wavelength is $\tau_{in}(\lambda)$. The pulse delay for the test fiber as a function of wavelength is $\tau_{out}(\lambda)$. The group delay $\tau(\lambda)$ per unit length at each wavelength is

$$\tau(\lambda) = \frac{\tau_{in}(\lambda) - \tau_{out}(\lambda)}{L_s - L_{ref}}$$

where L_s is the test sample fiber length in kilometers (km) and L_{ref} is the reference sample length in km.

The fiber chromatic dispersion is defined as the derivative, or slope, of the fiber group delay curve with respect to wavelength. Generally, the group delay as a function of wavelength is fit to a simple mathematical function and the derivative calculated. The range of wavelengths over which meaningful data is obtained depends on the wavelength range of optical source(s) used. The zero-dispersion wavelength (τ_0) and the zero-dispersion slope (S_0) are determined from the chromatic dispersion curve.

Fiber Geometry

End users perform fiber geometry measurements to reduce system attenuation and coupling loss resulting from poor fiber fabrication. Fiber attenuation and intrinsic coupling loss result from mismatches in the inherent fiber characteristics of two connecting fibers. Fiber mismatches occur when manufacturers fail to maintain optical or structural (geometrical) tolerances during the fiber fabrication process. Fiber geometry measurements performed in the laboratory identify fiber mismatches before the optical fiber is installed.

The procedure for measuring multimode and single mode fiber geometry is detailed in EIA/TIA-455-176. The fiber-geometrical parameters measured include cladding diameter, cladding noncircularity, core-cladding concentricity error, and core noncircularity. Figure 4.8 (Chapter 4) illustrates core noncircularity (ellipticity) and core-cladding concentricity error. The core noncircularity measurement is for multimode fibers only.

Other test methods are available for measuring other multimode and single mode fiber core parameters. Additional test methods exist for measuring multimode fiber core diameter and NA. For single mode fibers, the mode field diameter measurement replaces core diameter and NA measurements. Core diameter, numerical aperture, and mode field diameter measurements are identified and explained later in this chapter.

To make fiber geometry measurements, the input end of the fiber is overfilled and any cladding power stripped out. The output end of the fiber is prepared and viewed with a video camera. Generally the fiber is less than 10 m in length. An objective lens magnifies the output image (typically 20X) going to a video camera. The image from the video camera is displayed on a video monitor and is also sent to the computer for digital analysis.

The computer analyzes the image to identify the edges of the core and cladding. The centers r_c and r_g of the core and cladding, respectively, are found. The **cladding diameter** is defined as the average diameter of the cladding. The cladding diameter is twice the average radius (R_g). The **core diameter** is defined as the average diameter of the core. The core diameter is twice the average core radius (R_c).

Cladding noncircularity or ellipticity, is the difference between the smallest radius of the fiber (R_{gmin}) and the largest radius (R_{gmax}) divided by the average cladding radius (R_g). The value of the cladding noncircularity is expressed as a percentage.

The **core-cladding concentricity error** for multimode fibers is the distance between the core and cladding centers divided by the core diameter. Multimode core-cladding concentricity error is expressed as a percentage of core diameter. The core-cladding concentricity error for single mode fibers is defined as the distance between the core and cladding centers.

Core noncircularity is the difference between the smallest core radius (R_{cmin}) and the largest core radius (R_{cmax}) divided by the core radius (R_c). The value of core noncircularity is expressed as a percentage. Core noncircularity is measured on multimode fibers only.

Core Diameter

Core diameter is measured using EIA/TIA-455-58. The core diameter is defined from the refractive index profile $n(r)$ or the output near-field radiation pattern. Our discussion is limited to measuring the core diameter directly from the output near-field radiation pattern obtained using EIA/TIA-455-43.

The **near-field power distribution** is defined as the emitted power per unit area (radiance) for each position in the plane of the emitting surface. For this chapter, the emitting surface is the output area of a fiber-end face. Near-field power distributions describe the emitted power per unit area in the near-field region. The **near-field region** is the region close to the fiber-end face. In the near-field region, the distance between the fiber-end face and detector is in the micrometers (μm) range.

EIA/TIA-455-43 describes the procedure for measuring the near-field power distribution of optical waveguides. Output optics, such as lenses, magnify the fiber-end face and focus the fiber's image on a movable detector. Figure 5.7 shows an example setup for measuring the near-field power distribution. The image is scanned in a plane by the movable detector. The image may also be

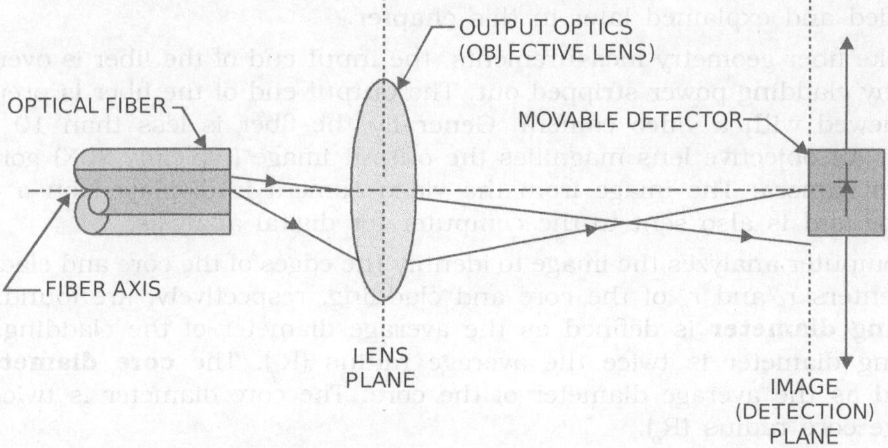

Fig. 5.7. *The measurement of the near-field power distribution.*

scanned by using a detector array. Detector arrays of known element size and spacing may provide a display of the power distribution on a video monitor. A record of the near-field power is kept as a function of scan position. The core diameter (D) is derived from the normalized output near-field radiation pattern. The normalized near-field pattern is plotted as a function of radial position on the fiber-end face. Figure 5.8 shows a plot of the normalized near-field radiation pattern as a function of scan position.

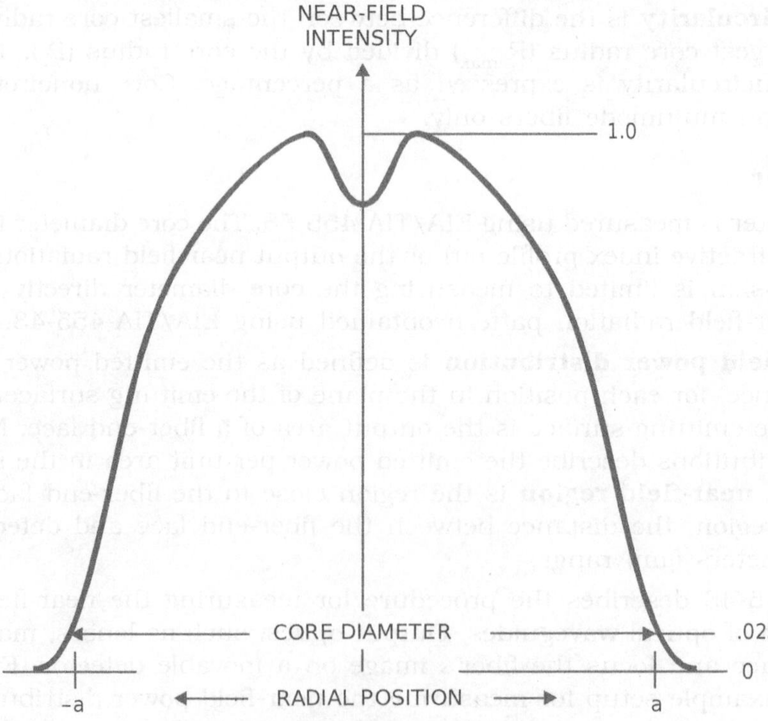

Fig. 5.8. *Near-field radiation pattern.*

The core diameter (D) is defined as the diameter at which the intensity is 2.5 percent of the maximum intensity (see Figure 5.8). The 2.5 percent points, or the 0.025 level, intersects the normalized curve at radial positions $-a$ and a. The core diameter is simply equal to 2a (D = 2a).

Numerical Aperture

The numerical aperture (NA) is a measurement of the ability of an optical fiber to capture light. The NA can be defined from the refractive index profile or the output far-field radiation pattern. Our discussion is limited to measuring the NA from the output far-field radiation pattern.

The NA of a multimode fiber having a near-parabolic refractive index profile is measured using EIA/TIA-455-177. In EIA/TIA-455-177, the fiber NA is measured from the output far-field radiation pattern. The far-field power distribution describes the emitted power per unit area in the far-field region. The **far-field region** is the region far from the fiber-end face. The **far-field power distribution** describes the emitted power per unit area as a function of angle θ some distance away from the fiber-end face. The distance between the fiber-end face and detector in the far-field region is in the centimeters (cm) range for multimode fibers and millimeters (mm) range for single mode fibers.

EIA/TIA-455-47 describes various procedures, or methods, for measuring the far-field power distribution of optical waveguides. These procedures involve either an angular or spacial scan. Figure 5.9 illustrates an angular and spacial scan for measuring the far-field power distribution.

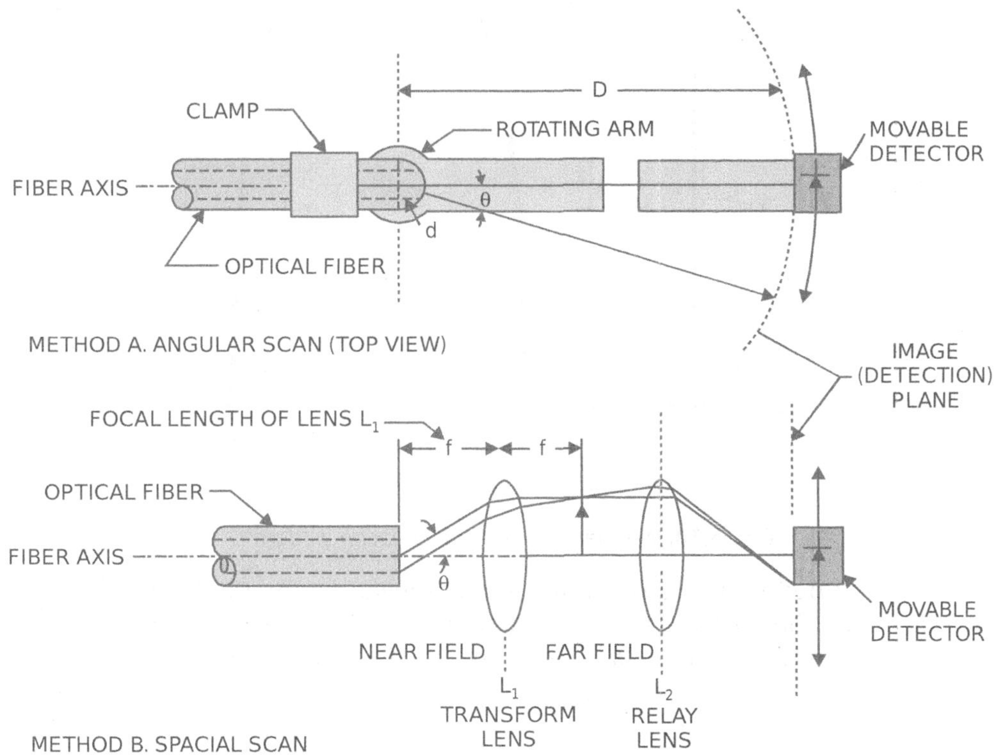

Fig. 5.9. *Angular and spacial scan methods for measuring the far-field power distribution.*

Figure 5.9 (method A) illustrates a far-field angular scan of the fiber-end face by a rotating detector. The fiber output radiation pattern is scanned by a rotating detector in the far-field. The detector rotates in a spherical manner. A record of the far-field power distribution is kept as a function of angle θ.

Figure 5.9 (method B) illustrates a far-field spacial scan of the fiber-end face by a movable (planar) detector. In a far-field spacial scan, lens L_1 performs a Fourier transform of the fiber output near-field pattern. A second lens, L_2, is positioned to magnify and relay the transformed image to the detector plane. The image is scanned in a plane by a movable detector. The scan position y in the Fourier transform plane is proportional to the far-field scan angle θ. A record of the far-field power distribution is kept as a function of the far-field scan angle.

The normalized far-field pattern is plotted as a function of the far-field scan angle θ. Figure 5.10 shows the plot of the normalized far-field radiation pattern as a function of scan angle.

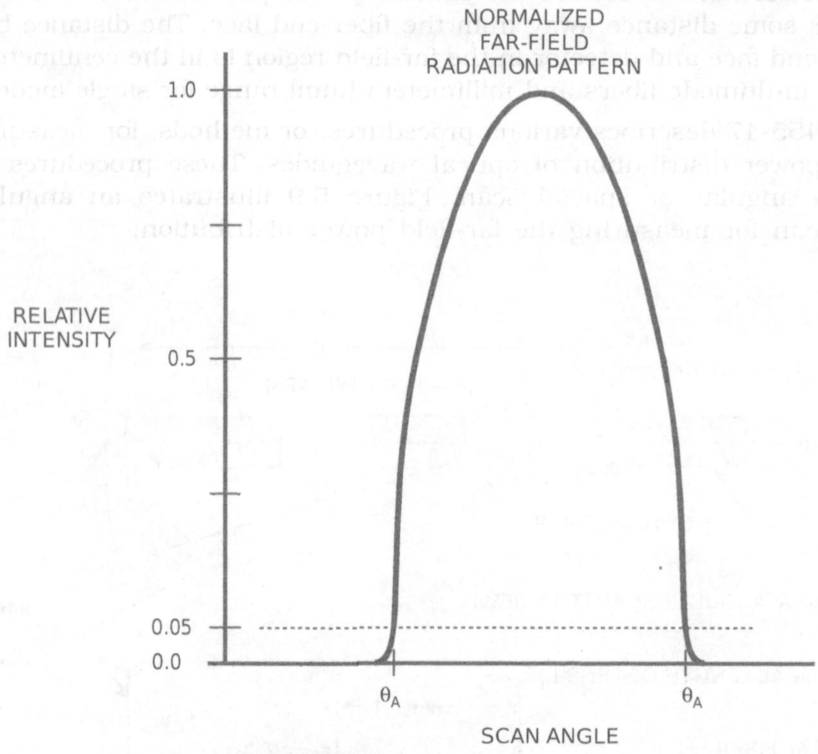

Fig. 5.10. *Normalized far-field radiation pattern.*

Fiber NA is defined by the 5 percent intensity level, or the 0.05 intensity level, as indicated in Fig. 5.10. The 0.05 intensity level intersects the normalized curve at scan angles θ_A and θ_B. The fiber NA is defined as

$$NA = \sin \theta_5$$

where θ_5 is the 5 percent intensity half angle. θ_5 is determined from θ_A and θ_B as shown below:

$$\theta_5 = \frac{\theta_A - \theta_B}{2}$$

Mode Field Diameter

The mode field diameter (MFD) of a single mode fiber is related to the spot size of the fundamental mode. This spot has a mode field radius w_0. The mode field diameter is equal to $2w_0$. The size of the mode field diameter correlates to the performance of the single mode fiber. Single mode fibers with large mode field diameters are more sensitive to fiber bending. Single mode fibers with small mode field diameters show higher coupling losses at connections.

The mode field diameter of a single mode fiber can be measured using EIA/TIA-455-167. This method involves measuring the output far-field power distribution of the single mode fiber using a set of apertures of various sizes. This far-field power distribution data is transformed into the near-field before using complex mathematical procedures. The mode field diameter is calculated from the transformed near field data. The mathematics behind the transformation between the far-field and near-field is too complicated for discussion in this chapter. Refer to EIA/TIA-455-167 for information on this transformation procedure.

Insertion Loss

Insertion loss is composed of the connection coupling loss and additional fiber losses in the fiber following the connection. In multimode fiber, fiber joints can increase fiber attenuation following the joint by disturbing the fiber's mode power distribution (MPD). Fiber joints may increase fiber attenuation because disturbing the MPD may excite radiative modes. Radiative modes are unbound modes that radiate out of the fiber contributing to joint loss. In single mode fibers, fiber joints can cause the second-order mode to propagate in the fiber following the joint. As long as the coupling loss of the connection is small, neither radiative modes (multimode fiber) or the second-order mode (single mode fiber) are excited.

Insertion loss of both multimode and single mode interconnection devices is measured using EIA/TIA-455-34. For Navy applications, an overfill launch condition is used at the input fiber. For other applications a mandrel wrap may be used to strip out high-order mode power. The length of fiber before the connection and after the connection may be specified for some applications. Power measurements are made on an optical fiber or fiber optic cable before the joint is inserted and after the joint is made. Figure 5.11 illustrates the mandrel wrap method of measuring the insertion loss of an interconnecting device in EIA/TIA-455-34.

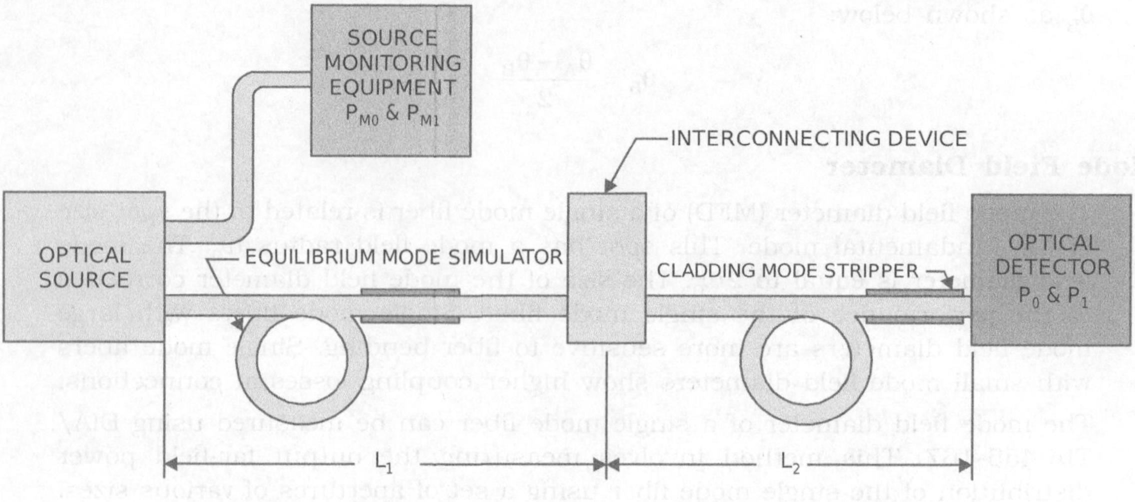

Fig. 5.11. *Insertion loss measurement of an interconnecting device.*

Initial power measurements at the detector (P_0) and at the source monitoring equipment (P_{M0}) are taken before inserting the interconnecting device into the test setup. The test fiber is then cut at the location specified by the end user. The cut results in a fiber of lengths L_1 and L_2 before and after the interconnection device that simulates the actual system configuration. After interconnection, the power at the detector (P_1) and at the source monitoring equipment (P_{M1}) is measured. The insertion loss is calculated as shown below:

$$\text{Insertion loss} = 10 \log\left\{\frac{P_1}{P_0} \times \frac{P_{M0}}{P_{M1}}\right\}$$

If the source power is constant, then the calculation of the insertion loss is similar to that of fiber attenuation.

Return Loss and Reflectance

Reflections occur at optical fiber connections. Optical power may be reflected back into the source fiber when connecting two optical fibers. In laser-based systems, reflected power reaching the optical source can reduce system performance by affecting the stability (operation) of the source. In addition, multiple reflections occur in fiber optic data links containing more than one connection. Multiple reflections can reduce data link performance by increasing the signal noise present at the optical detector.

Reflectance is a measure of the portion of incident light that is reflected back into the source fiber at the point of connection. Reflectance is given as a ratio (R) of the reflected light intensity to the incident light intensity. The reflectance ratio (R) for Fresnel reflection is discussed in Chapter 4.

Return loss and reflectance are measured using EIA/TIA-455-107. They are measured using an optical source connected to one input of a 2 × 2 fiber optic coupler. Light is launched into the component under test through the fiber optic coupler. The light reflected from the component under test is transmitted back through the fiber optic coupler to a detector connected to the other input port. The optical power is measured at the output of the device under test (P_o) and at the input port of the coupler where the detector is located (P_r). P_o is corrected to account for the loss in power through the device under test. P_r is corrected to account for the loss in power through the coupler and any other connection losses in the path. The reflectance is then given by the ratio P_r/P_o.

Return loss is the amount of loss of the reflected light compared with the power of the incident beam at the interface. The optical return loss at the fiber interface is defined as

$$\text{Return loss} = -10 \log R$$

Return loss is only the amount of optical power reflected and does not include power that is transmitted, absorbed, or scattered.

FIELD MEASUREMENTS

Field measurements differ from laboratory measurements because they measure the transmission properties of installed fiber optic components. Laboratory measurements can only attempt to simulate the actual operating conditions of installed components. Fiber optic component properties measured in the laboratory can change after the installation of these components on board ship. End users must perform field measurements to evaluate those properties most likely affected by the installation or repair of fiber optic components or systems.

The discussion on field measurements is limited to optical fiber and optical connection properties. Optical fiber and optical connection field measurements evaluate only the transmission properties affected by component or system installation or repair. Because optical fiber geometrical properties, such as core and cladding diameter and numerical aperture, are not expected to change, there is no need to remeasure these properties. The optical fiber properties that are likely to change include fiber attenuation (loss) and bandwidth. Bandwidth changes in the field tend to be beneficial, so field bandwidth measurement is generally not performed. If field bandwidth measurements are required, they are essentially the same as laboratory measurements so they will not be repeated. The optical connection properties that are likely to change are connection insertion loss and reflectance and return loss.

The installation and repair of fiber optic components on board ship can affect system operation. Microbends introduced during installation can increase fiber attenuation. Modal redistribution at fiber joints can increase fiber attenuation in the fiber after the joint. Fiber breaks or faults can prevent or severely disrupt system operation. Poor fiber connections can also increase insertion loss and degrade transmitter and receiver performance by increasing reflectance and return loss. End users should perform field measurements to verify that component performance is within allowable limits so system performance is not adversely affected.

There are additional differences in measuring optical fiber and optical connection properties in the field than in the laboratory. Field measurements require rugged, portable test equipment, unlike the sophisticated test equipment used in the laboratory. Field test equipment must provide accurate measurements in extreme environmental conditions. Since electrical power sources may not always be available in the field, test equipment should allow battery operation. In addition, while both fiber ends are available for conducting laboratory measurements, only one fiber end may be readily available for field measurements. Even if both fiber ends are available for field measurements, the fiber ends are normally located some distance apart. Therefore, field measurements may require two people.

The main field measurement technique involves optical time-domain reflectometry. An **optical time-domain reflectometer (OTDR)** is recommended for conducting field measurements on installed optical fibers or links of 50 meters or more in length. An OTDR requires access to only one fiber end. An OTDR measures the attenuation of installed optical fibers as a function of length. It also identifies and evaluates optical connection losses along a cable link and locates any fiber breaks or faults.

End users can also measure fiber attenuation and cable plant transmission loss using an optical power meter and a stabilized light source. End users use this measurement technique when optical time-domain reflectometry is not recommended. Measurements obtained with a stabilized light source and power meter are more accurate than those obtained with an OTDR. Measuring fiber attenuation and transmission loss using a power meter and light source requires access to both ends of the fiber or link. An **optical loss test set (OLTS)** combines the power meter and source functions into one physical unit.

Optical Time-Domain Reflectometry

End users use optical time-domain reflectometry to characterize optical fiber and optical connection properties in the field. In optical time-domain reflectometry, an **OTDR** transmits an optical pulse through an installed optical fiber. The OTDR measures the fraction of light that is reflected back due to Rayleigh scattering and Fresnel reflection. By comparing the amount of light scattered back at different times, the OTDR can determine fiber and connection losses. When several fibers are connected to form an installed cable plant, the OTDR can characterize optical fiber and optical connection properties along the entire length of the cable plant. A **fiber optic cable plant** consists of optical fiber cables, connectors, splices, mounting panels, jumper cables and other passive components. A cable plant does not include active components such as optical transmitters or receivers.

The OTDR displays the backscattered and reflected optical signal as a function of length. The OTDR plots half the power in decibels (dB) versus half the distance. Plotting half the power in dB and half the distance corrects for round trip effects. By analyzing the OTDR plot, or trace, end users can measure fiber attenuation and transmission loss between any two points along the cable plant. End users can also measure insertion loss and reflectance of any optical connection. In addition, end users use the OTDR trace to locate fiber breaks or faults.

Figure 5.12 shows an example OTDR trace of an installed cable plant. OTDR traces can have several common characteristics. An OTDR trace begins with an initial input pulse. This pulse is a result of Fresnel reflection occurring at the connection to the OTDR. Following this pulse, the OTDR trace is a gradual downsloping curve interrupted by abrupt shifts. Periods of gradual decline in the OTDR trace result from Rayleigh scattering as light travels along each fiber section of the cable plant. Periods of gradual decline are interrupted by abrupt shifts called point defects. A **point defect** is a temporary or permanent local deviation of the OTDR signal in the upward or downward direction. Point defects are caused by connectors, splices or breaks along the fiber length. Point defects or faults, can be reflective or nonreflective. An output pulse at the end of the OTDR trace indicates the end of the fiber cable plant. This output pulse results from Fresnel reflection occurring at the output fiber-end face.

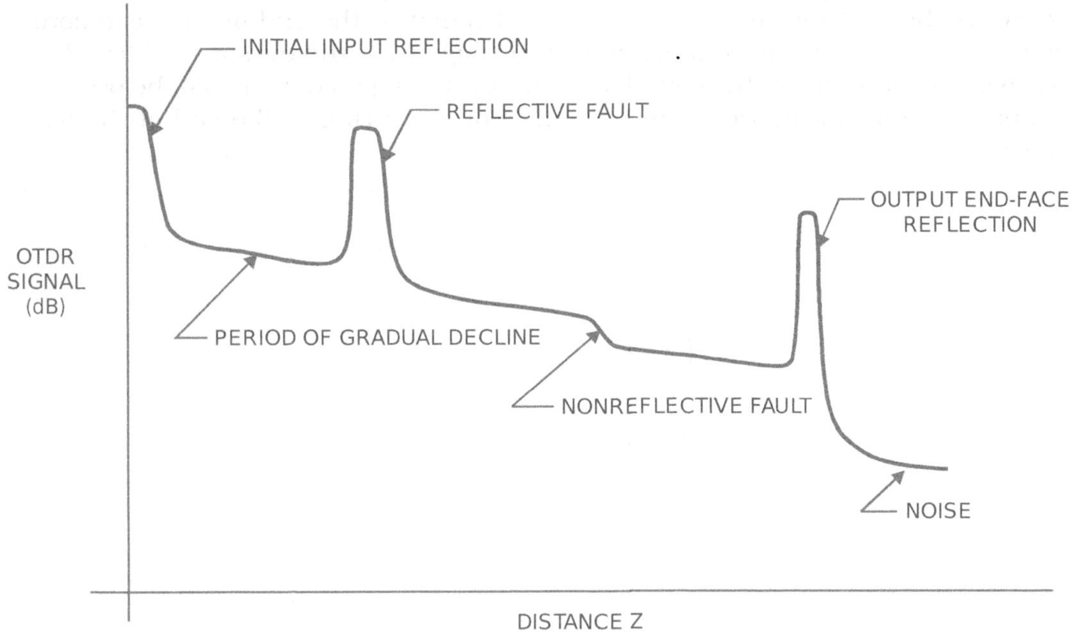

Fig. 5.12. *OTDR trace of an installed cable plant.*

Attenuation

The fiber optic test method for measuring the attenuation of an installed optical fiber using an OTDR is EIA/TIA-455-61. The accuracy of this test method depends on the user entering the appropriate source wavelength, pulse duration, and fiber length (test range) into the OTDR. In addition, the effective group index of the test fiber is required before the attenuation coefficient and accurate distances can be recorded. The group index (N) is provided by fiber manufacturers or is found using EIA/TIA-455-60. By entering correct test parameters, OTDR fiber attenuation values will closely coincide with those measured by the cutback technique.

Test personnel can connect the test fiber directly to the OTDR or to a **dead-zone** fiber. This dead-zone fiber is placed between the test fiber and OTDR to reduce the effect of the initial reflection at the OTDR on the fiber measurement. The dead-zone fiber is inserted because minimizing the reflection at a fiber joint is easier than reducing the reflection at the OTDR connection.

Fig. 5.13 illustrates the OTDR measurement points for measuring the attenuation of the test fiber using a dead-zone fiber. Fiber attenuation between two points along the test fiber is measured on gradual downsloping sections on the OTDR trace. There should be no point defects present along the portion of fiber being tested.

OTDRs are equipped with either manual or automatic cursors to locate points of interest along the trace. In Fig. 5.13, a cursor is positioned at a distance z_o on the rising edge of the reflection at the end of the dead-zone fiber. Cursors are also positioned at distances z_1 and z_2. The cursor positioned at z_1 is just beyond the recovery from the reflection at the end of the dead-zone fiber. Since no point defects are present in Fig. 5.13, the cursor positioned at z_2 locates the end of the test fiber. Cursor z_2 is positioned just before the output pulse resulting from Fresnel reflection occurring at the end of the test fiber.

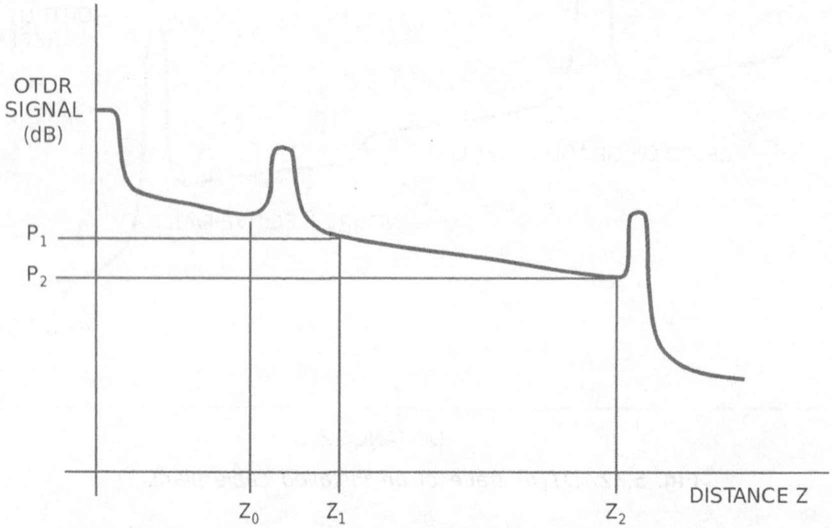

Fig. 5.13. *OTDR measurement points for measuring fiber attenuation using a dead-zone fiber.*

The attenuation of the test fiber between points z_1 and z_2 is $(P_1 - P_2)$ dB. The attenuation coefficient (α) is

$$\alpha = \frac{(P_1 - P_2)}{(z_2 - z_1)} \text{ dB/km}$$

The total attenuation of the fiber including the dead zone after the joint between the dead-zone fiber and test fiber is

$$\text{Attenuation} = (P_1 - P_2)\frac{(z_1 - z_0)}{(z_2 - z_1)} \text{ dB}$$

If fiber attenuation is measured without a dead-zone fiber, z_0 is equal to zero ($z_0 = 0$).

At any point along the length of fiber, attenuation values can change depending on the amount of optical power backscattered due to Rayleigh scattering. The amount of backscattered optical power at each point depends on the forward optical power and its **backscatter capture coefficient.** The backscatter capture coefficient varies with length depending on fiber properties. Fiber properties that may affect the backscatter coefficient include the refractive index profile, numerical aperture (multimode), and mode-field diameter (single mode) at the particular measurement point. The source wavelength and pulse width may also affect the amount of backscattered power.

By performing the OTDR attenuation measurement in each direction along the test fiber, test personnel can eliminate the effects of backscatter variations. Attenuation measurements made in the opposite direction at the same wavelength (within 5 nm) are averaged to reduce the effect of backscatter variations. This process is called bidirectional averaging. Bidirectional averaging is possible only if test personnel have access to both fiber ends. OTDR attenuation values obtained using bidirectional averaging should compare with those measured using the cutback technique in the laboratory.

Point Defects

Point defects are temporary or local deviations of the OTDR signal in the upward or downward direction. A point defect, or fault, can be reflective or nonreflective. A point defect normally exhibits a loss of optical power. However, a point defect may exhibit an apparent power gain. In some cases, a point defect can even exhibit no loss or gain. Refer back to Fig. 5.12; it illustrates a reflective fault and a nonreflective fault, both exhibiting loss. Figure 5.14 shows a nonreflective fault with apparent gain and a reflective fault with no apparent loss or gain.

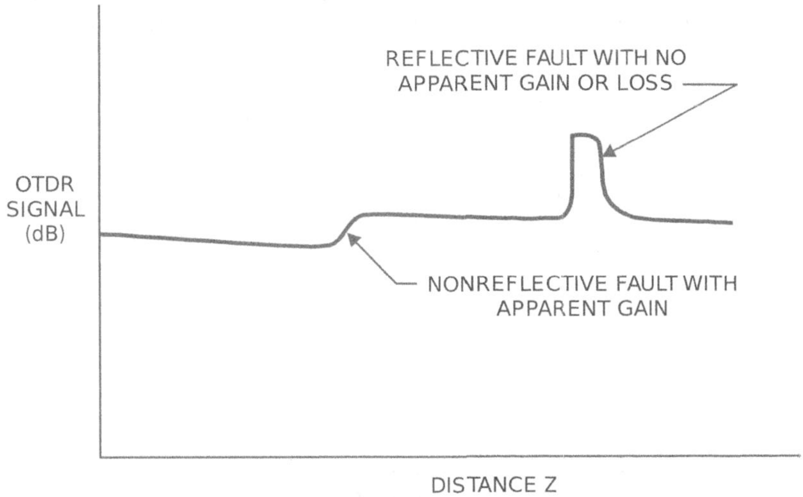

Figure 5.14. *An OTDR trace showing a nonreflective fault with apparent gain and a reflective fault with no apparent loss or gain.*

Point defects are located and measured using EIA/TIA-455-59. Test personnel must enter the appropriate input parameters including the source wavelength, the pulse duration, and the fiber or cable group index into the OTDR. The nature of fiber point defects depends on the value of each parameter entered by the end user. The pulse duration usually limits the length of the point defect while other input parameters, such as the wavelength, can vary its shape.

If the length of the fiber point defect changes with the pulse duration, then the OTDR signal deviation is in fact a point defect. If the length remains the same, then the OTDR signal deviation is a region of high fiber attenuation. Regions of high fiber attenuation are referred to as **attenuation non-uniformities.**

Fiber point defects occur from factory fiber splices or bends introduced during cable construction or installation. For shipboard applications, manufacturers are not allowed to splice fibers during cable construction. Fiber joints are natural sources of OTDR point defects. However, fiber breaks, cracks, or microbends introduced during cable installation are additional sources of point defects.

Point defects that occur at fiber joints are relatively easy to identify because the location of a fiber joint is generally known. A reflective or nonreflective fault occurs at a distance equal to fiber joint location. In most circumstances, an optical connector produces a reflective fault, while an optical splice produces a nonreflective fault.

Reflective and **nonreflective** faults occurring at distances other than fiber joint locations identify fiber breaks, cracks, or microbends. A fiber break produces a reflective fault because fiber breaks result in complete fiber separation. Fiber cracks and microbends generally produce nonreflective faults.

A point defect may exhibit apparent gain because the backscatter coefficient of the fiber present before the point defect is higher than that of the fiber present after. Test personnel measure the signal loss or gain by positioning a pair of cursors, one on each side of the point defect. Figure 5.15 illustrates

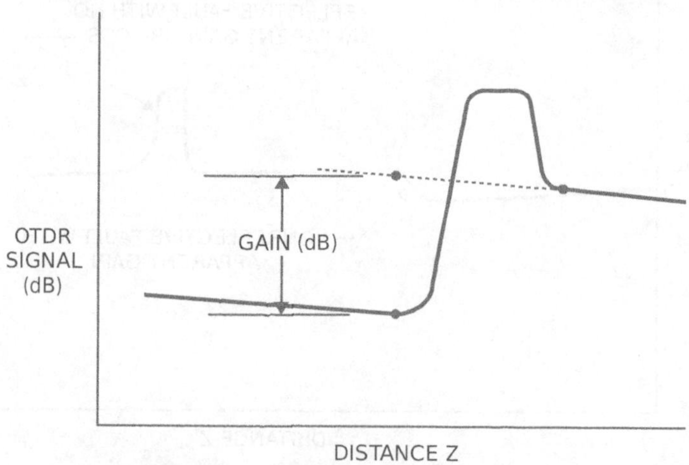

Fig. 5.15. *Extrapolation for a point defect showing an apparent signal gain.*

the positioning of the cursors for a point defect showing an apparent signal gain. The trace after the point defect is extrapolated as shown in Fig. 5.15. The vertical distance between the two lines in Fig. 5.15 is the apparent gain of the point defect.

Point defects exhibiting gain in one direction will exhibit an exaggerated loss in the opposite direction. Figure 5.16 shows the apparent loss shown by the OTDR for the same point defect shown in Fig. 5.15 when measured in the opposite direction. Bidirectional measurements are conducted to cancel the effects of backscatter coefficient variations. **Bidirectional averaging** combines the two values to identify the true signal loss. Bidirectional averaging is possible only if test personnel have access to both ends of the test sample.

OTDRs can also measure the return loss of a point defect. However, not all OTDRs are configured to make the measurement. To measure the return loss of a point defect, the cursors are placed in the same places as for measuring the loss of the point defect. The return loss of the point defect is displayed when the return loss option is selected on the OTDR. The steps for selecting the return loss option depend upon the OTDR being used.

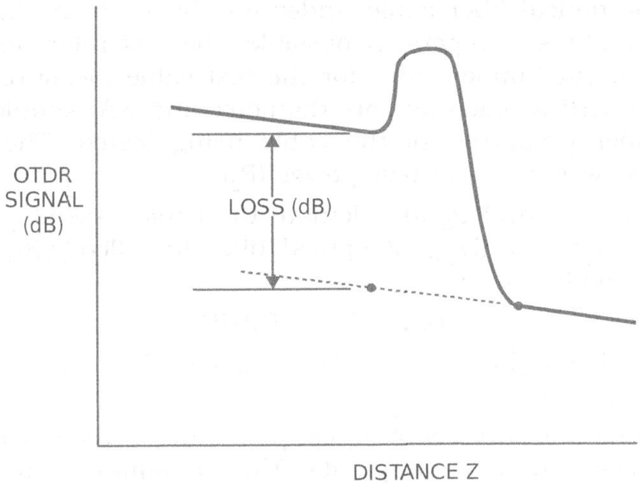

Fig. 5.16. *The exaggerated loss obtained at point defects exhibiting gain in one direction by conducting the OTDR measurement in the opposite direction.*

Power Meter

Test personnel also use an optical power meter and stabilized light source to measure fiber attenuation and transmission loss in the field. Optical power meter measurements are recommended when the length of an installed optical fiber cable or cable plant is less than 50 meters. A test jumper is used to couple light from the stabilized source to one end of the optical fiber (or cable plant) under test. An additional test jumper is also used to connect the other end of the optical fiber (or cable plant) under test to the power meter. Optical power meter measurements may be conducted using an **optical loss test set** (OLTS). An OLTS combines the power meter and source functions into one physical unit. When making measurements, it does not matter whether the stabilized source and power meter are in one physical unit or two.

Power meter measurements are conducted on individual optical fiber cables installed on board ship. The installed optical fiber cable must have connectors or terminations on both ends to make the measurement. If the installed optical fiber cable does not have connectors or terminations on both ends, an OTDR should be used to evaluate the cable. If the cable is too short for evaluation with an OTDR, cable continuity can be verified using a flashlight.

Power meter measurements for cable assembly link loss require that test personnel clean all optical connections at test jumper interfaces before performing any measurement. Test personnel should use cotton wipes dampened with alcohol to clean connectors and blow dry before making connections. End users should also ensure that test equipment calibration is current.

Power meter measurements connecting a test reference cable between the light source and power meter. The test reference cable has the same nominal fiber characteristics as the cable under test. The optical power present at the power meter is the reference power (P_1). Disconnect the test reference cable and connect the optical fiber cable under test between the light source and power meter using test jumpers. If possible, the test reference cable should be used as the input jumper cable for the test cable measurement. The test jumper fiber properties, such as core diameter and NA, should be nominally equal to the fiber properties of the cable being tested. The optical power present at the power meter is test power (P_2).

Test personnel use P_1 and P_2 to calculate the cable assembly link loss. The cable assembly link loss (B_{CA}) of optical fiber installed with connectors or terminations on both ends is

$$B_{CA} = (P_1 - P_2) \text{ dB}$$

The cable assembly link loss should always be less than the specified link loss for that particular link.

Besides measuring individual cables, test personnel measure the transmission loss of installed fiber optic cable plants. The transmission loss of fiber optic cable plants is measured using EIA/TIA-526-14 method B (multimode fiber) or EIA/TIA-526-7 (single mode fiber). The procedure measures the internal loss of the cable plant between points A and B, plus two connection losses. Figure 5.17 (A) illustrates the method described in EIA/TIA-526-14 method B for measuring the reference power (P_1). Figure 5.17 (B) shows the final test configuration for measuring the cable plant test power (P_2).

The procedure is exactly the same as described for measuring the link loss of an individual cable assembly. The total optical loss between any two termination points, including the end terminations, of the optical fiber cable plant link is measured. The measured cable plant link loss should always be less than the specified cable plant link loss.

Test personnel should conduct cable assembly link loss, and cable plant transmission loss measurements in both directions and at each system operational wavelength. By performing these measurements in each direction,

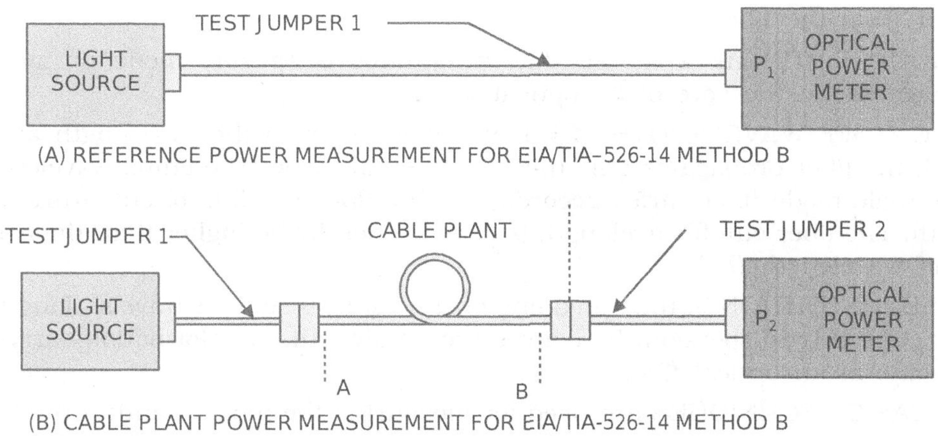

Fig. 5.17. *EIA/TIA-526-14 methods for measuring the reference power (P_1).*

test personnel can better characterize cable and link losses. Unlike optical time-domain reflectometry, bidirectional readings are always possible when performing power meter measurements. In power meter measurements, by definition, end users have access to both ends of the cable or cable plant.

SUMMARY

Now that you have completed this chapter, let's review some of the new terms, concepts, and ideas that you have learned. You should have a thorough understanding of these principles before moving on to Chapter 6.

END USERS (equipment manufacturers, shipbuilders, maintenance personnel, test personnel, and so on) should measure some component parameters upon receipt before installing the component into the fiber optic data link. In addition, they should measure some component parameters after installing or repairing fiber optic components in the field.

LABORATORY MEASUREMENTS of the optical fiber and optical connections performed by end users in the laboratory include attenuation, cutoff wavelength (single mode), bandwidth (multimode), chromatic dispersion, fiber geometry, core diameter, numerical aperture (multimode), mode field diameter (single mode), insertion loss, and reflectance and return loss.

ATTENUATION is the loss of optical power as light travels along the fiber. It is a result of absorption, scattering, bending, and other loss mechanisms.

The LAUNCH SPOT SIZE and the ANGULAR DISTRIBUTION **may affect multimode fiber attenuation measurement results by affecting modal distributions.**

An UNDERFILLED launch results when the launch spot size and angular distribution are smaller than that of the fiber core.

An OVERFILLED launch condition occurs when the launch spot size and angular distribution are larger than that of the fiber core.

A CLADDING-MODE STRIPPER is a device that removes any cladding mode power from the fiber.

A MODE FILTER is a device that attenuates a specific mode or modes propagating in the core of an optical fiber.

The CUTOFF WAVELENGTH of a single mode fiber is the wavelength above which the fiber propagates only the fundamental mode. The cutoff wavelength of a single mode fiber varies according to the fiber's radius of curvature and length. The fiber cutoff wavelength (λ_{cf}) will generally be higher than the cable cutoff wavelength (λ_{cc}).

PULSE DISTORTION is the spreading of the light pulse as it travels along the fiber caused by dispersion. It reduces the bandwidth, or information-carrying capacity, of an optical fiber.

Two BASIC TECHNIQUES are used for measuring the modal bandwidth of an optical fiber. The first characterizes dispersion by measuring the IMPULSE RESPONSE $h(t)$ of the fiber in the time domain. The second characterizes modal dispersion by measuring the BASEBAND **Frequency response H(f)** *of the fiber in the frequency domain.*

The LOWEST FREQUENCY at which the magnitude of the fiber Frequency response has decreased to one half its zero-frequency value is the –3 decibel (dB) optical power frequency (f_3dB).

CHROMATIC DISPERSION occurs because different colors of light travel through the fiber at different speeds. Since the different colors of light have different velocities, some colors arrive at the fiber end before others.

The DIFFERENTIAL GROUP DELAY $\tau(\lambda)$ is the variation in propagation delay that occurs because of the different group velocities of each wavelength in an optical fiber.

The RANGE OF WAVELENGTHS over which meaningful chromatic dispersion data is obtained depends on the wavelength range of optical source(s) used.

FIBER GEOMETRY MEASUREMENTS are performed by end users to reduce system attenuation and coupling loss resulting from poor fiber fabrication.

The CLADDING DIAMETER is the average diameter of the cladding.

CLADDING NONCIRCULARITY or ellipticity, is the difference between the smallest radius of the fiber (R_{gmin}) and the largest radius of the fiber (R_{gmax}) divided by the average cladding radius (R_g).

The CORE-CLADDING CONCENTRICITY ERROR for multimode fibers is the distance between the core and cladding centers divided by the core diameter. The core-cladding concentricity error for single mode fibers is defined as the distance between the core and cladding centers.

CORE NONCIRCULARITY is the difference between the smallest radius of the core (R_{cmin}) and the largest radius of the core (R_{cmax}) divided by the core radius (R_c).

The NEAR-FIELD POWER DISTRIBUTION is defined as the emitted power per unit area (radiance) for each position in the plane of the emitting surface.

The NEAR-FIELD REGION is the region close to the fiber-end face.

The CORE DIAMETER is derived from the normalized output near-field radiation pattern. The core diameter (D) is defined as the diameter at the 2.5 percent (0.025) level.

The NUMERICAL APERTURE (NA) is a measurement of the ability of a multimode optical fiber to capture light.

The FAR-FIELD POWER DISTRIBUTION describes the emitted power per unit area as a function of angle θ some distance away from the fiber-end face.

The FAR-FIELD REGION is the region far from the fiber-end face.

Single mode fibers with large MODE FIELD DIAMETERS are more sensitive to fiber bending. Single mode fibers with small mode field diameters show higher coupling losses at connections.

INSERTION LOSS is composed of the connection coupling loss and additional fiber losses in the fiber following the connection.

REFLECTANCE is a measure of the portion of incident light that is reflected back into the source fiber at the point of connection.

RETURN LOSS is the amount of loss of the reflected light compared with the power of the incident beam at the interface.

OPTICAL FIBER AND OPTICAL CONNECTION FIELD MEASUREMENTS measure only the transmission properties affected by component or system installation or repair.

OPTICAL TIME-DOMAIN REFLECTOMETRY is recommended for conducting field measurements on installed optical fibers or cable plants of 50 meters or more in length.

An OPTICAL LOSS TEST SET (OLTS) combines the power meter and source functions into one physical unit.

An OPTICAL TIME-DOMAIN REFLECTOMETER (OTDR) measures the fraction of light that is reflected back because of Rayleigh scattering and Fresnel reflection.

A FIBER OPTIC CABLE PLANT consists of optical fiber cables, connectors, splices, mounting panels, jumper cables, and other passive components. A cable plant does not include active components such as optical transmitters or receivers.

A POINT DEFECT is a temporary or permanent local deviation of the OTDR signal in the upward or downward direction. Point defects are caused by connectors, splices, or fiber breaks. Point defects, or faults, can be reflective or nonreflective.

A DEAD-ZONE fiber is placed between the test fiber and OTDR to reduce the influence of the initial pulse resulting from Fresnel reflection at the OTDR connection.

The AMOUNT OF OPTICAL POWER BACKSCATTERED because of Rayleigh scattering at one point depends on the forward optical power and the fibers backscatter capture coefficient.

The EFFECTS OF BACKSCATTER VARIATIONS can be eliminated by test personnel by performing the OTDR attenuation measurement in each direction along the test fiber and averaging (bidirectional readings).

A POINT DEFECT may exhibit apparent gain because the backscatter coefficient of the fiber present before the point defect is higher than that of the fiber present after.

To MEASURE FIBER ATTENUATION and TRANSMISSION LOSS in the field, test personnel use an optical power meter and stabilized light source.

OPTICAL POWER METER MEASUREMENTS are recommended when the length of an installed optical fiber cable or cable plant is less than 50 meters.

REVIEW QUESTIONS AND ANSWERS

Q. 1. *List the fiber geometrical measurements performed in the laboratory.*

Ans. Cladding diameter, core diameter, numerical aperture, and mode field diameter.

Q. 2. *End users measure the total attenuation of a fiber at the operating wavelength (λ). Write the equation for total attenuation (A), between an arbitrary point X and point Y located on an optical fiber.*

Ans. $A = 10 \log \frac{P_x}{P_y}$ dB.

Q. 3. *Will an optical fiber's attenuation coefficient vary with changes in wavelength?*

Ans. Yes.

Q. 4. *What two properties of the launch condition may affect multimode fiber attenuation measurements?*

Ans. Launch spot size and angular distribution.

Q. 5. *Does underfilling a multimode optical fiber excite mainly high-order or low-order modes?*

Ans. Low-order modes.

Q. 6. *Multimode optical fiber launch conditions are typically characterized as being overfilled or underfilled. Which of these optical launch conditions exists if the launch spot size and angular distribution are larger than that of the fiber core?*

Ans. Overfilled.

Q. 7. *A mode filter is a device that attenuates specific modes propagating in the core of an optical fiber. What mode propagating along single mode fibers do mode filters eliminate?*

Ans. Second-order mode.

Q. 8. *What are the two most common types of mode filters?*

Ans. Free-form loop and mandrel wrap.

Q. 9. *The cutoff wavelength of matched-clad and depressed-clad single mode fibers varies according to the fiber's radius of curvature and length. The cutoff wavelength of which single mode fiber type is more sensitive to length?*

Ans. Depressed-clad.

Q. 10. *Will the cutoff wavelength of uncabled fibers (λ_{cf}) generally have a value higher or lower than the cutoff wavelength of cabled fibers (λ_{cc})?*

Ans. Higher.

Q. 11. *Describe the –3 decibel (dB) optical power frequency (f_3dB).*

Ans. The –3 decibel (dB) is the lowest frequency at which the magnitude of the fiber Frequency response has decreased to one half its zero-frequency value.

Q. 12. *Delay differences between the source wavelengths occur as the optical signal propagates along the fiber. What is this called?*

Ans. Differential group delay $\tau(\lambda)$.

Q. 13. *What determines the range of wavelengths over which meaningful data is obtained for calculating the chromatic dispersion?*

Ans. The wavelength range of the optical source(s) used.

Q. 14. *Why do end users perform fiber geometry measurements in the laboratory?*

Ans. To reduce system attenuation and coupling loss resulting from poor fiber fabrication.

Q. 15. *Define cladding diameter.*

Ans. The cladding diameter is the average diameter of the cladding.

Q. 16. *Explain the difference between multimode and single mode core-cladding concentricity errors.*

Ans. Multimode core-cladding concentricity error is the distance between the core and cladding centers expressed as a percentage of core diameter while the single mode core-cladding concentricity error is just the distance between the core and cladding centers.

Q. 17. *Near-field power distributions describe the emitted power per unit area in the near-field region. Describe the near-field region.*

Ans. The near-field region is the region close to the fiber-end face.

Q. 18. *How is the core diameter defined?*

Ans. The core diameter is defined as the diameter at which the near-field intensity is 2.5 percent of the maximum intensity.

Q. 19. *Far-field power distributions describe the emitted power per unit area as a function of angle θ in the far-field region. Describe the far-field region.*

Ans. The far-field region is the region far from the fiber-end face.

Q. 20. *Will fiber coupling loss generally increase or decrease if the mode field diameter of a single mode fiber is decreased?*

Ans. Increase.

Q. 21. *In multimode fibers, how do fiber joints increase fiber attenuation following the joint?*

Ans. By disturbing the fiber's mode power distribution (MPD).

Q. 22. *List two effects that reflections can have on a fiber optic data link.*

Ans. Reduce the stability of the system source and increase the signal noise present at the optical detector.

Q. 23. *Reflectance is given as what ratio?*

Ans. The ratio of reflected optical power to incident optical power.

Q. 24. *Does return loss include power that is transmitted, absorbed, and/or scattered?*

Ans. No.

Q. 25. *Is it essential for end users to remeasure optical fiber geometrical properties after installation in the field?*

Ans. No.

Q. 26. *When is an OTDR recommended for conducting field measurements on installed optical.*

Ans. When installed optical fiber cables or links are 50 meters or more in length.

Q. 27. *An OTDR measures the fraction of light that is reflected back from the fiber or link under test. What causes light to be reflected back into the OTDR?*

Ans. Rayleigh scattering and Fresnel reflection.

Q. 28. *List the types of fiber optic components considered part of a fiber optic cable plant.*

Ans. Optical fiber cables, connectors, splices, mounting panels, jumper cables, and other passive components.

Q. 29. *What is a temporary or permanent local deviation of the OTDR signal in the upward or downward direction called?*

Ans. A point defect.

Q. 30. *Why is a dead-zone fiber placed between the test fiber and OTDR when conducting attenuation measurements?*

Ans. To reduce the effect of the initial reflection at the OTDR.

Q. 31. *The amount of backscattered optical power at each point depends on what two properties?*

Ans. Forward optical power and backscatter capture coefficient.

Q. 32. *How can test personnel eliminate the effects of backscatter variations?*

Ans. By performing the OTDR attenuation measurements in each direction along the test fiber.

Q. 33. *If the length of the fiber point defect changes with pulse duration, is the OTDR signal deviation a point defect or a region of high fiber attenuation?*

Ans. A point defect.

Q. 34. *Give the type of fault (reflective or nonreflective) normally produced by: (a) fiber breaks, (b) fiber cracks, and (c) fiber microbends.*

Ans. (*a*) Reflective, (*b*) nonreflective, and (*c*) nonreflective.

Q. 35. *Explain how a point defect may exhibit an apparent gain.*

Ans. A point defect may exhibit apparent gain because the backscatter coefficient of the fiber present before the point defect is higher than that of the fiber present after.

Q. 36. *A point defect exhibiting an apparent gain in one direction will exhibit what when measured in the opposite direction?*

Ans. An exaggerated loss.

Q. 37. *When is an optical power meter measurement recommended for conducting field measurements on installed optical fiber cables or cable plants?*

Ans. When an installed optical fiber cable or cable plant is less than 50 meters in length.

Q. 38. *If an installed optical fiber cable does not have connectors or terminations on both ends, how should the cable be tested?*

Ans. With an OTDR unless it is less than 50 meters in length. If it is less than 50 meters in length, continuity should be verified with a flashlight.

6 OPTICAL SOURCES AND FIBER OPTIC TRANSMITTERS

Learning Objectives

Upon completion of this chapter, you should be able to do the following:

- Explain the principal properties of an optical source and fiber optic transmitter.
- Discuss the optical emission properties of semiconductor light-emitting diodes (LEDs) and laser diodes (LDs).
- Describe the operational differences between surface-emitting LEDs (SLEDs), edge-emitting LEDs (ELEDs), superluminescent diodes (SLDs), and laser diodes.
- Describe typical fiber optic transmitter packages.

INTRODUCTION TO OPTICAL SOURCES AND FIBER OPTIC TRANSMITTERS

Chapter 1 taught you that a fiber optic data link has three basic functions.

One function is that a fiber optic data link must convert an electrical signal to an optical signal permitting the transfer of data along an optical fiber.

The fiber optic device responsible for that signal conversion is a fiber optic transmitter.

A **fiber optic transmitter** is a hybrid device.

It converts electrical signals into optical signals and launches the optical signals into an optical fiber. A fiber optic transmitter consists of an interface circuit, a source drive circuit and an optical source. The interface circuit accepts the incoming electrical signal and processes it to make it compatible with the source drive circuit. The source drive circuit intensity modulates the optical source by varying the current through the source.

An **optical source** converts electrical energy (current) into optical energy (light). Light emitted by an optical source is launched or coupled, into an optical fiber for transmission. Fiber optic data link performance depends on the amount of optical power (light) launched into the optical fiber. This chapter attempts to provide an understanding of light-generating mechanisms within the main types of optical sources used in fiber optics.

OPTICAL SOURCE PROPERTIES

The development of efficient semiconductor optical sources, along with low-loss optical fibers, led to substantial improvements in fiber optic communications. Semiconductor optical sources have the physical characteristics and performance properties necessary for successful implementations of fiber optic systems. It is desirable that optical sources:

- Be compatible in size to low-loss optical fibers by having a small light-emitting area capable of launching light into fiber.
- Launch sufficient optical power into the optical fiber to overcome fiber attenuation and connection losses allowing for signal detection at the receiver.
- Emit light at wavelengths that minimize optical fiber loss and dispersion.
- Optical sources should have a narrow spectral width to minimize dispersion.
- Allow for direct modulation of optical output power.

Maintain stable operation in changing environmental conditions (such as temperature). Cost less and be more reliable than electrical devices, permitting fiber optic communication systems to compete with conventional systems semiconductor optical sources suitable for fiber optic systems range from inexpensive LEDs to more expensive semiconductor lasers. Semiconductor LEDs and laser diodes LDs are the principal light sources used in fiber optics.

OPERATING WAVELENGTH

Fiber optic communication systems operate in the 850 nm, the 1300 nm and the 1550 nm wavelength windows. Semiconductor sources are designed to operate at wavelengths that minimize optical fiber absorption and maximize system bandwidth. By designing an optical source to operate at specific wavelengths, absorption from impurities in the optical fiber, such as hydroxyl ions (OH^-), can be minimized. Maximizing system bandwidth involves designing optical fibers and sources that minimize chromatic and intermodal dispersion at the intended operational wavelength.

Initially, the material properties of semiconductor optical sources provided for optical emission in the 850 nm wavelength region. An 850 nm operational wavelength avoids fiber absorption loss from OH^- impurities near the 900 nm wavelength. Light sources for 850 nm systems were originally semiconductor LEDs and lasers. Currently, most 850 nm systems use LEDs as a light source. LEDs operating at 850 nm provide sufficient optical power for short-distance, low-bandwidth systems. However, multimode fiber dispersion, the relatively high fiber attenuation, and the LED's relatively low optical output power prevent the use of these devices in longer-distance, higher bandwidth systems.

The first development allowing the operational wavelength to move from 850 nm to 1300 nm was the introduction of multimode graded-index fibers. Multimode graded-index fibers have substantially lower intermodal dispersion than multimode step-index fibers. Systems operating at 850 nm cannot take full advantage of the fiber's low intermodal dispersion because of high chromatic dispersion at 850 nm. However, the use of multimode graded-index fibers allow 850 nm LEDs to operate satisfactorily in short-distance, higher bandwidth systems.

Following the enhancements in multimode fiber design, next generation LEDs were designed to provide optical emission in the 1300 nm region. Multimode graded-index fiber systems using these LEDs can operate over longer distances and at higher bandwidths than 850 nm systems. Longer distances and higher bandwidths are possible because fiber material losses and dispersion are significantly reduced at the 1300 nm region.

Advances in single mode fiber design and construction speed the development of semiconductor LEDs and LDs optimized for single mode fibers. Single mode fibers have very low dispersion values. However, existing LEDs were unable to focus and launch sufficient optical power into single mode fibers for long-haul, very high-bandwidth communication systems. New semiconductor LEDs and LDs capable of operating with single mode fibers at 1300 nm were developed to take advantage of single mode fiber's very low value of dispersion. Additionally, LEDs and LDs operating at 1550 nm were developed to take advantage of the fiber's lowest loss.

SEMICONDUCTOR LIGHT-EMITTING DIODES AND LASER DIODES

Semiconductor LEDs Emit Incoherent Light

Spontaneous emission of light in semiconductor LEDs produces light waves that lack a fixed-phase relationship. Light waves that lack a fixed-phase relationship are referred to as **incoherent light.** Spontaneous emission of light is discussed in more detail later in this chapter. The use of LEDs in single mode systems is severely limited because they emit unfocused incoherent light. Even LEDs developed for single mode systems are unable to launch sufficient optical power into single mode fibers for many applications. LEDs are the preferred optical source for multimode systems because they can launch sufficient power at a lower cost than semiconductor LDs.

Semiconductor LDs Emit Coherent Light

LDs produce light waves with a fixed-phase relationship (both spatial and temporal) between points on the electromagnetic wave. Light waves having a fixed-phase relationship are referred to as coherent light. Stimulated emission of light is discussed later in this chapter. Since semiconductor LDs emit more focused light than LEDs, they are able to launch optical power into both single mode and multimode optical fibers. However, LDs are usually used only in single mode fiber systems because they require more complex driver circuitry and cost more than LEDs.

Optical power produced by optical sources can range from microwatts (μW) for LEDs to tens of milliwatts (mW) for semiconductor LDs. However, it is not possible to effectively couple all the available optical power into the optical fiber for transmission.

The amount of optical power coupled into the fiber is the relevant optical power. It depends on the following factors:

- The angles over which the light is emitted.
- The size of the source's light-emitting area relative to the fiber core size.
- The alignment of the source and fiber.
- The coupling characteristics of the fiber (such as the NA and the refractive index profile).

Typically, semiconductor lasers emit light spread out over an angle of 10 to 15 degrees. Semiconductor LEDs emit light spread out at even larger angles. Coupling losses of several decibels can easily occur when coupling light from an optical source to a fiber, especially with LEDs.

Source-to-fiber coupling efficiency is a measure of the relevant optical power. The coupling efficiency depends on the type of fiber that is attached to the optical source. Coupling efficiency also depends on the coupling technique.

Source-to-fiber coupling involves centering a flat fiber-end face over the emitting region of the light source. If the fiber end face is directly placed over the source emitting region, it is referred to as **butt coupling.** If the source's output light pattern is larger than the fiber's acceptance pattern, source-to-fiber coupling efficiency may be improved by placing a small lens between the source and fiber. Lensing schemes improve coupling efficiency when coupling both LEDs and LDs to optical fibers.

SEMICONDUCTOR MATERIAL AND DEVICE OPERATING PRINCIPLES

Understanding optical emission in semiconductor lasers and LEDs requires knowledge of semiconductor material and device properties. Providing a complete description of semiconductor properties is beyond the scope of this introductory manual. In this chapter we only discuss the general properties of semiconductor LEDs and LDs.

Semiconductor sources are diodes, with all of the characteristics typical of diodes. However, their construction includes a special layer, called the active layer, which emits photons (light particles) when a current passes through the layer. The particular properties of the semiconductor are determined by the materials used and the layering of the materials within the semiconductor. **Silicon (Si)** and **gallium arsenide (GaAs)** are the two most common semiconductor materials used in electronic and electro-optic devices. In some cases other elements, such as aluminum (Al), indium (In) and phosphorus (P), are added to the base semiconductor material to modify the semiconductor properties. These elements are called **dopants.**

Current flowing through a semiconductor optical source causes it to produce light. An in-depth description of either of the two processes by which this occurs is beyond the scope of this module. However, we discuss elementary descriptions in the following paragraphs.

LEDs generally produce light through **spontaneous emission** when a current is passed through them. Spontaneous emission is the random generation of photons within the active layer of the LED. The emitted photons move in random directions. Only a certain percentage of the photons exit the semiconductor and are coupled into the fiber. Many of the photons are absorbed by the LED materials and the energy dissipated as heat.

This process causes the light output from an LED to be incoherent, have a broad spectral width, and have a wide output pattern.

Laser diodes are much more complex than LEDs. Laser is an acronym for light amplification by the stimulated emission of radiation. Laser diodes produce light through stimulated emission when a current is passed through them. **Stimulated emission** describes how light is produced in any type of laser. In the laser diode, photons, initially produced by spontaneous emission interact with the laser material to produce additional photons. This process occurs within the active area of the diode called the laser cavity. The process does not affect the original photon. The stimulated photon has many of the same properties (wavelength, direction, phase) as the original photon.

As with the LED, not all of the photons produced are emitted from the laser diode. Some of the photons are absorbed and the energy dissipated as heat. The emission process and the physical characteristics of the diode cause the light output to be coherent, have a narrow spectral width, and have a narrow output pattern.

It is important to note that in both LED and laser diodes all of the electrical energy is not converted into optical energy. A substantial portion is converted to heat. Different LED and laser diode structures convert differing amounts of electrical energy into optical energy.

LIGHT-EMITTING DIODES

A **light-emitting diode** (LED) is a semiconductor device that emits incoherent light, through spontaneous emission, when a current is passed through it. Typically LEDs for the 850 nm region are fabricated using GaAs and AlGaAs. LEDs for the 1300 nm and 1550 nm regions are fabricated using InGaAsP and InP.

The basic LED types used for fiber optic communication systems are the surface-emitting LED (SLED), the edge-emitting LED (ELED) and the superluminescent diode (SLD). LED performance differences help link designers decide which device is appropriate for the intended application. For short-distance (0 to 3 km), low-data-rate fiber optic systems, SLEDs and ELEDs are the preferred optical source. Typically, SLEDs operate efficiently for bit rates up to 250 megabits per second (Mb/s). Because SLEDs emit light over a wide area (wide far-field angle), they are almost exclusively used in multimode systems.

For medium-distance, medium-data-rate systems, ELEDs are preferred.

ELEDs may be modulated at rates up to 400 Mb/s. ELEDs may be used for both single mode and multimode fiber systems. Both SLDs and ELEDs are used in long-distance, high-data-rate systems. SLDs are ELED-based diodes designed to operate in the superluminescence mode. A further discussion on superluminescence is provided later in this chapter. SLDs may be modulated at bit rates of over 400 Mb/s.

Surface-Emitting LEDs

The surface-emitting LED (shown in Fig. 6.1) is also known as the Burrus LED in honor of C. A. Burrus, its developer. In SLEDs, the size of the primary active region is limited to a small circular area of 20 μm to 50 μm in diameter. The active region is the portion of the LED where photons are emitted. The primary active region is below the surface of the semiconductor substrate perpendicular to the axis of the fiber.

A **well** is etched into the substrate to allow direct coupling of the emitted light to the optical fiber. The etched well allows the optical fiber to come into close contact with the emitting surface.

In addition, the epoxy resin that binds the optical fiber to the SLED reduces the refractive index mismatch, increasing coupling efficiency.

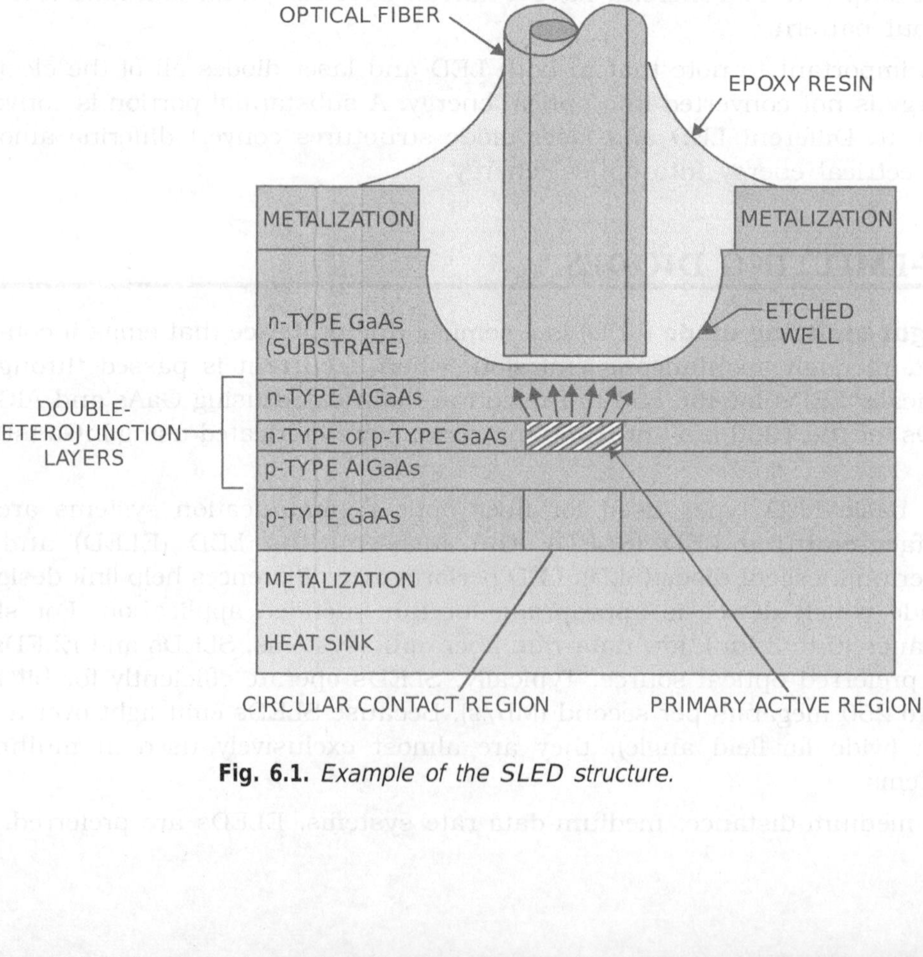

Fig. 6.1. *Example of the SLED structure.*

Edge-Emitting LEDs

The demand for optical sources for longer distance, higher bandwidth systems operating at longer wavelengths led to the development of edge-emitting **LEDs.** Figure 6.2 shows a typical ELED structure. It shows the different layers of semiconductor material used in the ELED. The primary active region of the ELED is a narrow stripe, which lies below the surface of the semiconductor substrate. The semiconductor substrate is cut or polished so that the stripe runs between the front and back of the device.

The polished or cut surfaces at each end of the stripe are called **facets.**

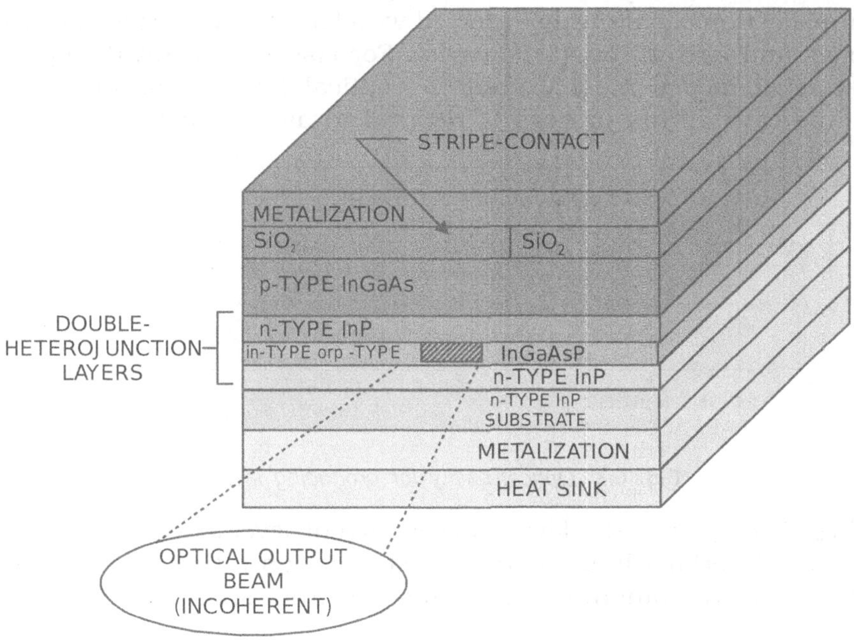

Fig. 6.2. *Example of the ELED structure.*

In an ELED the rear facet is highly reflective and the front facet is antireflection-coated. The rear facet reflects the light propagating toward the rear end-face back toward the front facet. By coating the front facet with antireflection material, the front facet reduces optical feedback and allows light emission. ELEDs emit light only through the front facet. ELEDs emit light in a narrow emission angle allowing for better source-to-fiber coupling. They couple more power into small NA fibers than SLEDs. ELEDs can couple enough power into single mode fibers for some applications. ELEDs emit power over a narrower spectral range than SLEDs. However, ELEDs typically are more sensitive to temperature fluctuations than SLEDs.

LASER DIODES

A **laser** is a device that produces optical radiation by the process of stimulated emission. It is necessary to contain photons produced by stimulated emission within the laser active region.

Figure 6.3 shows an optical cavity formed to contain the emitted photons by placing one reflecting mirror at each end of an amplifying medium. One mirror is made partially reflecting so that some radiation can escape from the cavity for coupling to an optical fiber.

Only a portion of the optical radiation is amplified. For a particular laser structure, there are only certain wavelengths that will be amplified by that laser. Amplification occurs when selected wavelengths, also called laser modes, reflect back and forth through the cavity. For lasing to occur, the optical gain of the selected modes must exceed the optical loss during one round-trip through the cavity. This process is referred to as optical feedback.

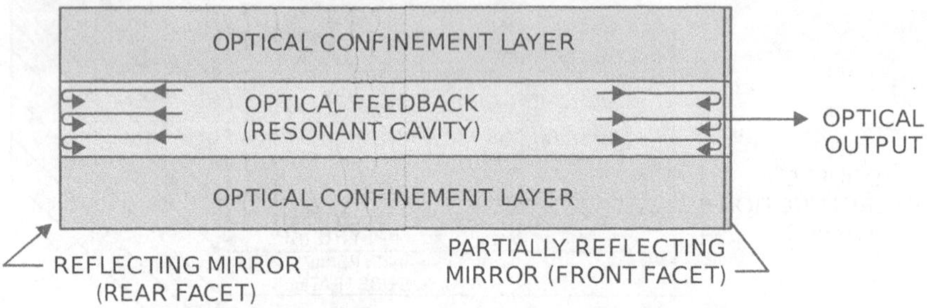

Fig. 6.3. *Optical cavity for producing lasing.*

The **lasing threshold** is the lowest drive current level at which the output of the laser results primarily from stimulated emission rather than spontaneous emission. Figure 6.4 illustrates the transition from spontaneous emission to

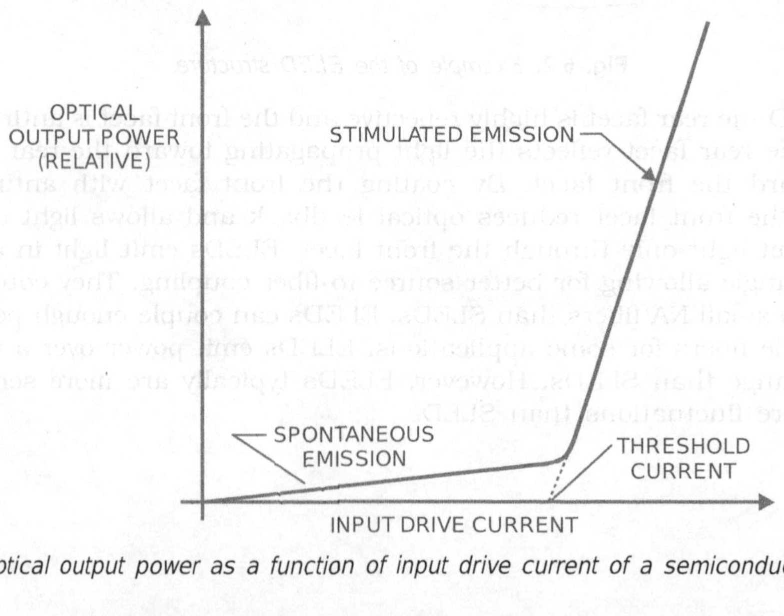

Fig. 6.4. *The optical output power as a function of input drive current of a semiconductor laser diode.*

stimulated emission by plotting the relative optical output power and input drive current of a semiconductor laser diode. The lowest current at which stimulated emission exceeds spontaneous emission is the **threshold current.**

Before the threshold current is reached, the optical output power increases only slightly with small increases in drive current. However, after the threshold current is reached, the optical output power increases significantly with small changes in drive currents.

Many types of materials including gas, liquid, and semiconductors can form the lasing medium. However, in this chapter we only discuss semiconductor laser diodes. Semiconductor laser diodes are the primary lasers used in fiber optics. A laser diode emits light that is highly monochromatic and very directional.

This means that the LD's output has a narrow spectral width and small output beam angle.

A semiconductor LD's geometry is similar to an ELED with light-guiding regions surrounding the active region. Optical feedback is established by making the front facet partially reflective. This chapter provides no diagram detailing LD structures because they are similar to ELEDs in design. The rear facet is typically coated with a reflective layer so that all of the light striking the facet is reflected back into the active region. The front facet is typically left uncoated so that most of the light is emitted. By increasing the drive current, the diode becomes a laser.

At currents below the threshold current, LDs function as ELEDs.

To optimize Frequency response, laser diodes are often biased above this laser threshold. As a result, in an LD fiber optic system, light is modulated between a high power level and a lower power level, but never shut off. LDs typically can be modulated at frequencies up to over 2 gigahertz (GHz). Some lasers are capable of being modulated at frequencies over 20 GHz.

There are several important differences between LDs and LEDs. One is that LEDs usually lack reflective facets and in some cases are designed to suppress reflections back into the active region. Another is that lasers tend to operate at higher drive currents to produce light. A higher driver current results in more complicated drive circuits and more heat dissipation in the device.

LDs are also much more temperature sensitive than either SLEDs or ELEDs. Increases in the laser temperature significantly reduce laser output power. Increases in laser temperature beyond certain limits result in the loss of lasing. When lasers are used in many applications, the temperature of the laser must be controlled. Typically, electronic coolers, called **thermo-electric (TE)** coolers, are used to cool LDs in system applications.

SUPERLUMINESCENT DIODES

Superluminescence occurs when the spontaneous emissions of an ELED experience gain due to higher injected currents and reflections from facets. Superluminescent diodes (SLDs) are differentiated from both conventional LEDs and LDs. Although the output is not fully coherent.

SLDs emit light that consists of amplified spontaneous emissions. The spectral width and beam angle of SLDs are narrower than that of conventional LEDs and wider than that of LDs.

An SLD is, in essence, a combination of a laser and an ELED. SLDs are similar in geometry to lasers but have no built-in optical feedback mechanism required by laser diodes for stimulated emission to achieve lasing. SLDs have structural features similar to those of ELEDs that suppress the lasing action by reducing the reflectivity of the facets. SLDs are essentially highly optimized ELEDs.

While SLDs operate like ELEDs at low current levels, their output power increases superlinearly and the spectral width narrows at high currents.

Optical gain resulting from the higher injection currents causes the superlinear power increase and narrowing of the spectral width.

The advantages of SLDs over conventional LEDs include higher coupled power, narrower spectral width, and greater bandwidths. The disadvantages include nonlinear power-current characteristics, higher temperature sensitivity, and lower reliability.

FIBER OPTIC TRANSMITTERS

As stated previously, a fiber optic transmitter is a hybrid electro-optic device. It converts electrical signals into optical signals and launches the optical signals into an optical fiber. A fiber optic transmitter consists of an interface circuit, a source drive circuit, and an optical source. The interface circuit accepts the incoming electrical signal and processes it to make it compatible with the source drive circuit. The source drive circuit intensity modulates the optical source by varying the current through it.

The optical signal is coupled into an optical fiber through the transmitter output interface.

Although semiconductor LEDs and LDs have many similarities, unique transmitter designs result from differences between LED and LD sources. Transmitter designs compensate for differences in optical output power, response time, linearity, and thermal behaviour between LEDs and LDs to ensure proper system operation. Nonlinearities caused by junction heating in LEDs and mode instabilities in LDs necessitate the use of linearizing circuits within the transmitter in some cases.

Fiber optic transmitters using LDs require more complex circuitry than transmitters using LEDs. The basic requirement for digital systems is for drive circuitry to switch the optical output on and off at high speeds in response to logic voltage levels at the input of the source drive circuit.

Because LDs are threshold devices, LDs are supplied with a bias just below the threshold in the off state. This bias is often referred to as **prebias.** One reason for prebiasing the LD is to reduce the turn-on delay in digital systems. Most LD transmitters contain output **power control circuitry** to compensate for temperature sensitivity. This circuitry maintains the LD output at a constant average value by adjusting the bias current of the laser. In most cases LED transmitters do not contain output power control circuitry. LD and LED transmitters may also contain cooling devices to maintain the source at a relatively constant temperature. Most LD transmitters either have an internal **thermo electric cooler** or require a relatively controlled external temperature.

Because LDs require more complex circuitry than LEDs, fiber optic transmitters using LDs are more expensive.

Transmitter output interfaces generally fall into two categories: optical connectors and optical fiber pigtails. **Optical pigtails** are attached to the transmitter optical source. This pigtail is generally routed out of the transmitter package as a coated fiber in a loose buffer tube or a single fiber cable. The pigtail is either soldered or epoxied to the transmitter package to provide fiber strain relief. The buffer tube or single fiber cable is also attached to the transmitter package to provide additional strain relief.

The transmitter output interface may consist of a fiber **optical connector.** The optical source may couple to the output optical connector through an intermediate optical fiber. One end of the optical fiber is attached to the source. The other end terminates in the transmitter optical output connector. The optical source may also couple to the output optical connector without an intermediate optical fiber. The optical source is placed within the transmitter package to launch power directly into the fiber of the mating optical connector. In some cases **lenses** are used to more efficiently couple light from the source into the mating optical connector.

FIBER OPTIC TRANSMITTER PACKAGES

Fiber optic transmitters come in various sizes and shapes. The least complex fiber optic transmitters are typically packaged in **transistor outline (TO) cans** or hybrid microcircuit modules in **dual inline packages (DIPs).** These simple transmitters may require separate circuitry in the system equipment to provide an acceptable input signal to the transmitter. More complex fiber optic transmitters are available that have some or all of the signal conditioning circuitry integrated into the package. These transmitters typically are packaged in hybrid microcircuit modules in either DIP or butterfly lead packages, circuit cards or complete stand-alone fiber optic converters. Stand-alone fiber optic converters and circuit cards generally contain sources in either TO cans or one of the hybrid microcircuit packages. For commercial applications, the most popular transmitter packages are the TO can and the DIP hybrid microcircuit.

FIBER OPTIC TRANSMITTER APPLICATIONS

Fiber optic transmitters can be classified into two categories: **digital** and **analog.** Digital transmitters produce two discrete optical power levels. These levels are essentially on and off with the exception that some light is emitted in the off state by some transmitters. Analog transmitters continuously vary the output optical power level as a function of the input electrical signal.

Digital Applications

Different types of fiber optic transmitters are used for different digital applications. For each specific application, the link data rate, transmission length, and operating environment influence the source type, center wavelength, spectral width, and package type chosen.

For **low-data rate applications,** fiber optic transmitters generally use LEDs operating in either the 850 nm or 1300 nm window as their source. For the lowest data rates (0 to 20 megabits per second (Mbps)), sources tend to operate in the 850 nm window. For **moderate data rates** (50 to 200 Mbps), sources tend to operate in the 1300 nm window. Laser sources are almost never used in low-data-rate applications. Laser sources are only used when extremely high transmitter output powers are required in the application. The packages found in low-data-rate applications include all of the package types discussed earlier.

For **high-data rate applications,** most fiber optic transmitters use laser diodes as sources. The sources typically operate in either the 1300 nm or 1550 nm windows. Most high-data-rate applications use LDs as the optical source and operate in the 1300 nm region. Almost all 1550 nm systems use an LD as the optical source. 1550 nm transmitters are usually only used in the extremely long distance high-data-rate applications (undersea links, etc.). High-data-rate transmitters are generally hybrid microcircuit modules or complete circuit cards. Almost all high-data-rate transmitters contain power control circuitry. Depending upon the application, high-data-rate transmitters may contain TE coolers.

Analog Applications

Different types of fiber optic transmitters are also used for different analog applications. For each specific application, analog signal type, transmission length, and operating environment influence the source type, center wavelength, spectral width, and package type chosen.

For **low-frequency applications,** analog fiber optic transmitters generally use LEDs operating in either the 850 nm or 1300 nm window. Typical low frequency applications are analog audio and single channel video systems. For these systems, sources tend to operate in the 850 nm window. For **moderate frequency applications,** sources tend to operate in the 1300 nm window. These types of systems include multi-channel analog audio and video systems as well as frequency modulated (FM) systems.

Laser sources are almost never used in low- or moderate-frequency analog applications. The main reason for this is the added circuit complexity that laser sources require. Laser sources are only used if extremely high transmitter output powers are required in the application. Most low-frequency analog transmitters are hybrid microcircuit modules, circuit cards or stand-alone boxes.

For **high-frequency applications,** analog fiber optic transmitters use laser diodes as sources. Typical high frequency applications are cable television trunk line and raw radar remoting applications. The LDs typically operate in either the 1300 nm or 1550 nm windows. 1550 nm transmitters are typically used in cable television trunk line applications. Other applications may use either 1300 nm or 1550 nm LDs. High frequency transmitters are predominately circuit cards, but some hybrid microcircuit modules are also used. All high frequency analog transmitters contain TE coolers as well as linearization and power control circuitry.

SUMMARY

Now that you have completed this chapter, let's review some of the new terms, concepts and ideas that you have learned. You should have a thorough understanding of these principles before moving on to Chapter 7.

A FIBER OPTIC TRANSMITTER is a hybrid electro-optic device. It converts electrical signals into optical signals and launches the optical signals into an optical fiber.

An OPTICAL SOURCE converts electrical energy (current) into optical energy (light).

The principal LIGHT SOURCES used in fiber optics are semiconductor LEDs and LDs.

SEMICONDUCTOR LD's emit coherent light. Light waves having a fixed-phase relationship are referred to as coherent light.

SEMICONDUCTOR LED's emit incoherent light. Light waves that lack a fixed-phase relationship are referred to as incoherent light.

The RELEVANT OPTICAL POWER is the amount of optical power coupled into the fiber. It depends on the angle over which the light is emitted, the size of the source's light-emitting area relative to the fiber core size, the alignment of the source and fiber, and the coupling characteristics of the fiber (such as the NA and the refractive index profile).

SOURCE-TO-FIBER COUPLING EFFICIENCY is a measure of the relevant optical power.

SILICON (Si) and GALLIUM ARSENIDE (GaAs) are the two most common semiconductor materials used in electronic and electro-optic devices. In a semiconductor device, PHOTONS (LIGHT) are emitted when current flows through the active area.

SPONTANEOUS EMISSION occurs when photons are emitted in a random manner. Spontaneous emission produces incoherent light.

STIMULATED EMISSION occurs when a photon interacts with the laser material to produce additional photons.

A LIGHT-EMITTING DIODE is a semiconductor device that emits incoherent light, through spontaneous emission, when a current is passed through it. The basic LED types used for fiber optic communication systems are the

SURFACE-EMITTING LED (SLED), the EDGE-EMITTING LED (ELED), and the SUPERLUMINESCENT DIODE (SLD).

In SURFACE-EMITTING LED's (SLEDs), the size of the primary active region is limited to a small circular area of 20 m to 50 m in diameter. The active region is the portion of the LED where photons are emitted. SLEDs usually emit more total power into the air gap at the fiber interface than an ELED, but they do not launch as much power into the fiber. SLEDS also tend to emit power over a wider spectral range than ELED.

EDGE-EMITTING LED's (ELEDs) emit light in a narrow emission angle allowing for better source-to-fiber coupling. They couple more power into small NA fibers than SLEDs. The polished or cut surfaces at each end of the ELED active stripe are called FACETS.

SUPERLUMINESCENCE occurs when the spontaneous emissions of an ELED experience gain due to higher injected currents and reflections from facets.

SUPERLUMINESCENT DIODES (SLDs) are similar in geometry to lasers but have no built-in optical feedback mechanism required by laser diodes for stimulated emission to achieve lasing. Although the output is not fully coherent, superluminescent diodes (SLDs) emit light that consists of amplified spontaneous emissions. The spectral width and beam angle of SLDs are narrower than that of conventional LEDs and wider than that of LDs.

The ADVANTAGES of SLDs over conventional LEDs include higher coupled power, narrower spectral width, and greater bandwidths. The DISADVANTAGES include nonlinear power-current characteristics, higher temperature sensitivity, and lower reliability.

A LASER is a device that produces optical radiation using stimulated emission rather than spontaneous emission. Laser is an acronym for light amplification by the stimulated emission of radiation.

The LASING THRESHOLD is the lowest drive level at which the output of the laser results primarily from stimulated emission rather than spontaneous emission.

The THRESHOLD CURRENT is the lowest current at which stimulated emission exceeds spontaneous emission.

A LASER DIODE is a semiconductor diode that emits coherent light by lasing. The LD's output has a narrow spectral width and small output beam angle.

TRANSMITTER OUTPUT INTERFACES fall into two categories: optical connectors and optical fiber pigtails.

FIBER OPTIC TRANSMITTERS using LDs require more complex circuitry than transmitters using LEDs.

Because LDs are threshold devices, LDs are supplied with a bias just below the threshold in the off state. This bias is often referred to as a prebias.

The least complex FIBER OPTIC TRANSMITTERS are typically packaged in transistor outline (TO) cans or hybrid microcircuit modules in dual inline packages (DIPs).

More complex FIBER OPTIC TRANSMITTERS typically are packaged in hybrid microcircuit modules in either DIP or butterfly lead packages, circuit cards, or complete stand-alone fiber optic converters.

FIBER OPTIC TRANSMITTERS can be classified into two categories: digital and analog.

DIGITAL TRANSMITTERS modulate the fiber optic source between two discrete optical power levels. These levels are essentially on and off with the exception that some light is emitted in the off state by some transmitters.

ANALOG TRANSMITTERS continuously vary the output optical power level as a function of the input electrical signal.

For LOW-DATA-RATE APPLICATIONS (0 to 20 Mbps), fiber optic transmitters generally use LEDs operating in either the 850 nm or 1300 nm window.

For MODERATE-DATA-RATE APPLICATIONS (50 to 200 Mbps), fiber optic transmitters generally use LEDs operating in the 1300 nm window.

For HIGH-DATA-RATE APPLICATIONS, most fiber optic transmitters use laser diodes as sources.

LASER SOURCES are almost never used in low- or moderate-frequency analog applications because LED sources require much less complex circuitry.

REVIEW QUESTIONS AND ANSWERS

Q. 1. *What are the three parts of a fiber optic transmitter?*

Ans. Interface circuit, source drive circuit, and an optical source.

Q. 2. *Which part of a fiber optic transmitter converts the processed electrical signal to an optical signal?*

Ans. The optical source.

Q. 3. *LEDs operating at 850 nm provide sufficient optical power for short-distance, low-bandwidth multimode systems. List three conditions that prevent the use of LEDs in longer distance, higher bandwidth multimode systems.*

Ans. Multimode fiber dispersion, the relatively high fiber attenuation, and the LED's relatively low optical output power.

Q. 4. *Why can multimode graded-index fiber 1300 nm systems using LEDs operate over longer distances and at higher bandwidths than 850 nm systems?*

Ans. Longer distances and higher bandwidths are possible because fiber material losses and dispersion are significantly reduced at the 1300 nm region.

Q. 5. *Semiconductor LEDs emit incoherent light. Define incoherent light.*

Ans. Light waves that lack a fixed-phase relationship.

Q. 6. *Which semiconductor sources (LD or LED) emit more focused light and are capable of launching sufficient optical power into both single mode and multimode fibers?*

Ans. LDs.

Q. 7. *The amount of optical power coupled into an optical fiber depends on what four factors?*

Ans. (1) The angles over which the light is emitted. (2) The size of the source's light-emitting area relative to the fiber core size. (3) The alignment of the source and fiber. (4) The coupling characteristics of the fiber (such as the NA and the refractive index profile).

Q. 8. *What are the two most common semiconductor materials used in electronic and electro-optic devices?*

Ans. Silicon and gallium arsenide.

Q. 9. *What is a laser?*

Ans. A laser is a device that produces optical radiation using stimulated emission rather than spontaneous emission.

Q. 10. *Describe stimulated emission.*

Ans. A photon initially produced by spontaneous emission in the active region interacts with the laser material to produce additional photons.

Q. 11. *What are the three basic LED types?*

Ans. Surface-emitting LEDs (SLEDs), edge-emitting LEDs (ELEDs) and super luminescent diodes (SLDs).

Q. 12. *Which types of LEDs are the preferred optical sources for short-distance, low-data-rate fiber optic systems?*

Ans. SLEDs and ELEDs.

Q. 13. *What are facets?*

Ans. Cut or polished surfaces at each end of the narrow active region of an ELED.

Q. 14. *What is lowest current at which stimulated emission exceeds spontaneous emission in a semiconductor laser called?*

Ans. Threshold current.

Q. 15. *Describe the output of a laser diode.*

Ans. The LD's output has a narrow spectral width and small output beam angle.

Q. 16. *Which type of optical source usually lacks reflective facets and in some cases are designed to suppress reflections back into the active region?*

Ans. LED.

Q. 17. *Which type of optical source tends to operate at higher drive currents to produce light?*

Ans. Laser.

Q. 18. *Are the effects of temperature changes on LDs more or less significant than for LEDs?*

Ans. More.

Q. 19. *Specify the mechanism that SLDs lack that is required by laser diodes to achieve lasing.*

Ans. SLDs have no built-in optical feedback mechanism.

Q. 20. *How does the source drive circuit intensity modulate the source?*

Ans. By varying the current through the source.

Q. 21. *What is a prebias?*

Ans. A current applied in the laser off state just less than the threshold current.

Q. 22. *Is the drive circuitry generally more complex for an LED or a laser diode? Why?*

Ans. For a laser diode. The laser diode transmitter generally contains output power control circuitry and may contain a TE cooler and some circuitry associated with the TE cooler.

Q. 23. *What are the two types of output interfaces for fiber optic transmitters?*

Ans. Optical connectors and optical fiber pigtails.

Q. 24. *List five common fiber optic transmitter packages.*

Ans. TO can, DIP, butterfly lead microcircuits, circuit cards, and stand-alone optical fiber converters.

Q. 25. *What type of source is typically used in low-data-rate digital applications?*

Ans. LED.

Q. 26. *Why would a laser diode be used in a low-data-rate digital application?*

Ans. When extremely high transmitter output powers are required.

Q. 27. *What type of source is generally used in high-data-rate digital applications?*

Ans. Laser diode.

Q. 28. *Why are LEDs preferred over laser diodes for low- and moderate-frequency analog applications?*

Ans. LEDs require less complex circuitry than lasers.

■■■

7 OPTICAL DETECTORS AND FIBER OPTIC RECEIVERS

Learning Objectives

Upon completion of this chapter, you should be able to do the following:

- Explain the principal properties of an optical detector and fiber optic receiver.
- Detail semiconductor optical detector performance and capability requirements necessary for the successful implementation of fiber optic systems.
- List the main components of a fiber optic receiver.
- Discuss receiver sensitivity, dynamic range, and other key operational parameters used to define receiver performance.

INTRODUCTION TO OPTICAL DETECTORS AND FIBER OPTIC RECEIVERS

Chapter 6 taught you that a fiber optic transmitter is an electro-optic device capable of accepting electrical signals, converting them into optical signals, and launching the optical signals into an optical fiber. The optical signals propagating in the fiber become weakened and distorted because of scattering, absorption, and dispersion. The fiber optic device responsible for converting the weakened and distorted optical signal back to an electrical signal is a fiber optic receiver.

A **fiber optic receiver** is an electro-optic device that accepts optical signals from an optical fiber and converts them into electrical signals. A typical fiber optic receiver consists of an optical detector, a low-noise amplifier, and other circuitry used to produce the output electrical signal (see Fig. 7.1). The optical detector converts the incoming optical signal into an electrical signal. The amplifier then amplifies the electrical signal to a level suitable for further signal processing. The type of other circuitry contained within the receiver depends on what type of modulation is used and the receiver electrical output requirements.

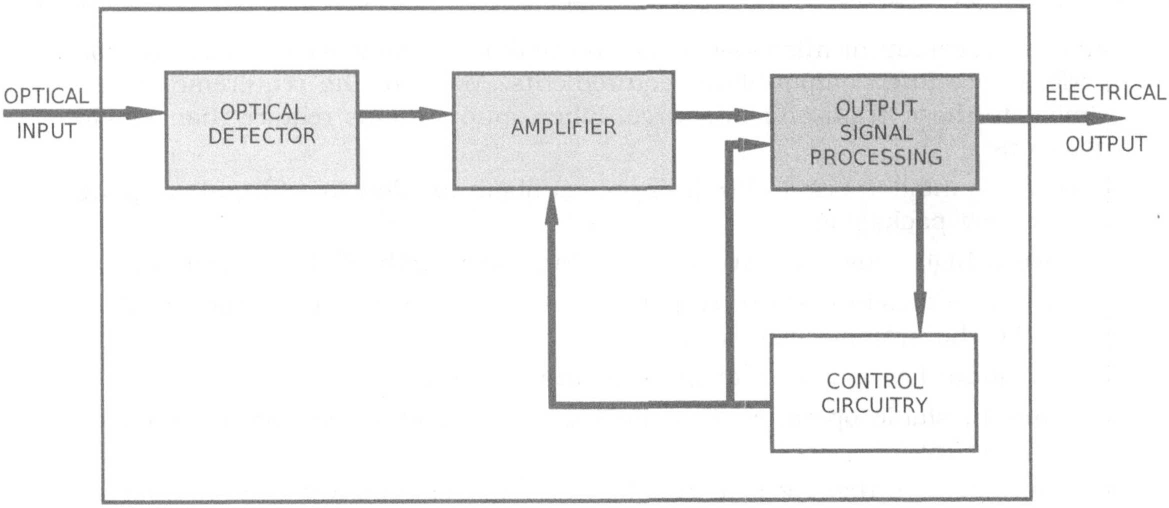

Fig. 7.1. *Block diagram of a typical fiber optic receiver.*

Receiver spectral response, sensitivity, frequency response, and dynamic range are key receiver performance parameters that can affect overall system operation. The choice of optical detector materials and structures determines the spectral response. Silicon (Si), gallium arsenide (GaAs), and gallium aluminum arsenide (GaAlAs) are typical detector materials used for receiver operation in the 850 nm wavelength region. germanium (Ge), indium phosphide (InP) and indium gallium arsenide (InGaAs) are examples of detector materials used for receiver operation in the 1300 nm and 1550 nm wavelength regions.

The **receiver sensitivity** is the minimum amount of optical power required to achieve a specific receiver performance.

For digital transmission at a given data rate and coding, this performance is described by a maximum bit-error rate (BER). In analog systems, for a given modulation and bandwidth, it is described by a minimum signal-to-noise ratio (SNR). **Dynamic range** refers to the range of optical power levels over which the receiver operates within the specified values. It usually is described by the ratio of the maximum input power to the sensitivity. Before discussing receiver sensitivity, bandwidth, dynamic range, and frequency response in more detail, we discuss the main types of optical detectors used in fiber optics.

OPTICAL DETECTORS

A **transducer** is a device that converts input energy of one form into output energy of another. An **optical detector** is a transducer that converts an optical signal into an electrical signal. It does this by generating an electrical current proportional to the intensity of incident optical radiation. The relationship between the input optical radiation and the output electrical current is given by the detector responsivity. Responsivity is discussed later in this chapter.

OPTICAL DETECTOR PROPERTIES

Fiber optic communications systems require that optical detectors meet specific performance and compatibility requirements. Many of the requirements are similar to those of an optical source. Fiber optic systems require that optical detectors:

- Be compatible in size to low-loss optical fibers to allow for efficient coupling and easy packaging.
- Have a high sensitivity at the operating wavelength of the optical source.
- Have a sufficiently short response time (sufficiently wide bandwidth) to handle the system's data rate.
- Contribute low amounts of noise to the system.
- Maintain stable operation in changing environmental conditions, such as temperature.

Optical detectors that meet many of these requirements and are suitable for fiber optic systems are semiconductor photodiodes. The principal optical detectors used in fiber optic systems include semiconductor positive-intrinsic-negative (PIN) photodiodes and avalanche photodiodes (APDs).

SEMICONDUCTOR PHOTODIODES

Semiconductor photodiodes generate a current when they absorb photons (light). The amount of current generated depends on the following factors:

- The wavelengths of the incident light and the responsivity of the photodiode at those wavelengths
- The size of the photodiode active area relative to the fiber core size
- The alignment of the fiber and the photodiode

The optical fiber is coupled to semiconductor photodiodes similarly to the way optical sources are coupled to optical fibers. Fiber-to-photodiode coupling involves centering the flat fiber-end face over the photodiode active area. This is normally done directly by butt coupling the fiber up to the photodiode surface. As long as the photodiode active area is larger than that of the fiber core, fiber-to-detector coupling losses are very low. In some cases a lens may be used to couple the fiber end-face to the detector. However, this is not typically done.

SEMICONDUCTOR MATERIAL AND DEVICE PROPERTIES

The mechanism by which optical detectors convert optical power into electrical current requires knowledge of semiconductor material and device properties. As stated in Chapter 6, providing a complete description of these properties is beyond the scope of this manual. In this chapter we only discuss the general properties of semiconductor PINs and APDs.

Semiconductor detectors are designed so that optical energy (photons) incident on the detector active area produces a current. This current is called a **photocurrent.** The particular properties of the semiconductor are determined by the materials used and the layering of the materials within the device. Silicon (Si), gallium arsenide (GaAs), germanium (Ge), and indium phosphide (InP) are the most common semiconductor materials used in optical detectors. In some cases aluminum (Al) and indium (In) are used as dopants in the base semiconductor material.

Responsivity

Responsivity is the ratio of the optical detector's output photocurrent in amperes to the incident optical power in watts. The responsivity of a detector is a function of the wavelength of the incident light and the efficiency of the device in responding to that wavelength. For a particular material, only photons of certain wavelengths will generate a photocurrent when they are absorbed. Additionally, the detector material absorbs some wavelengths better than others. These two properties cause the wavelength dependence in the detector responsivity. Responsivity is a useful parameter for characterizing detector performance because it relates the photocurrent generated to the incident optical power.

PIN PHOTODIODES

A **PIN photodiode** is a semiconductor positive-negative (p-n) structure with an intrinsic region sandwiched between the other two regions (see Fig. 7.2). It is normally operated by applying a reverse-bias voltage. The magnitude of the reverse-bias voltage depends on the photodiode application, but typically is less than a few volts. When no light is incident on the photodiode, a current is still produced. This current is called the **dark current.**

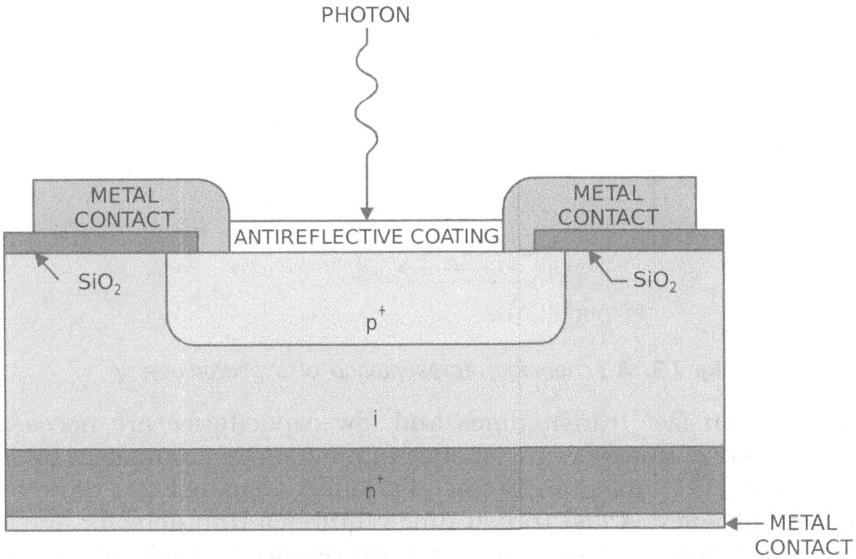

Fig. 7.2. *The basic structure of a PIN photodiode.*

The dark current is the leakage current that flows when a reverse bias is applied and no light is incident on the photodiode. Dark current is dependent on temperature. While dark current may initially be low, it will increase as the device temperature increases.

Response Time

There are several factors that influence the response time of a photodiode and its output circuitry (see Fig. 7.3).

The most important of these are the thickness of the detector active area and the detector RC time constant. The detector thickness is related to the amount of time required for the electrons generated to flow out of the detector active area. This time is referred to as the electron **transit time.** The thicker the detector active area, the longer the transit time will be.

The **capacitance (C)** of the photodiode and the **resistance (R)** of the load form the RC time constant. The capacitance of the photodetector must be kept small to prevent the RC time constant from limiting the response time. The photodiode capacitance consists mainly of the junction capacitance and any capacitance relating to packaging. The **RC time constant** is given by t_{RC} = RC.

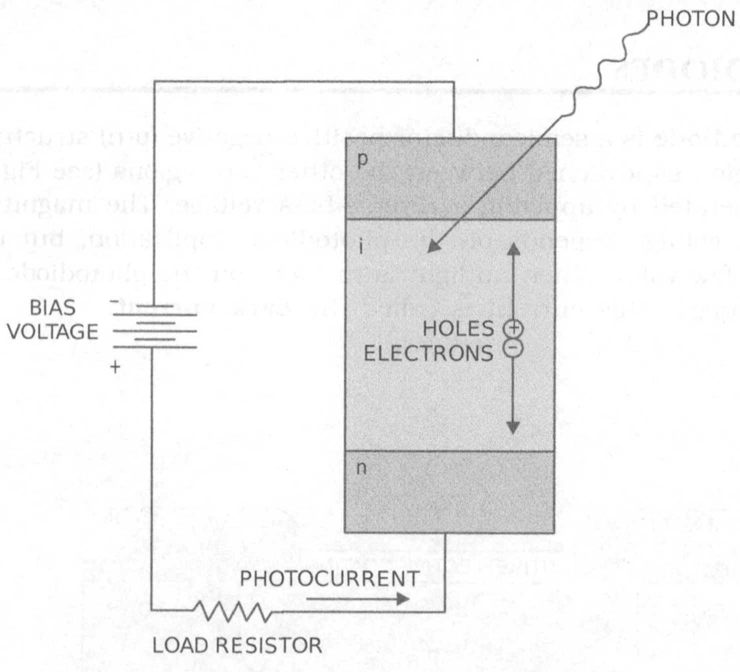

Fig. 7.3. *A schematic representation of a photodiode.*

Trade-offs between fast transit times and low capacitance are necessary for high-speed response. However, any change in photodiode parameters to optimize the transit time and capacitance can also affect responsivity, dark current, and coupling efficiency. A fast transit time requires a thin detector active area, while low capacitance and high responsivity require a thick active region.

The diameter of the detector active area can also be minimized. This reduces the detector dark current and minimizes junction capacitance. However, a minimum limit on this active area exists to provide for efficient fiber-to-detector coupling.

Linearity

Reverse-biased photodetectors are highly linear devices. Detector **linearity** means that the output electrical current (photocurrent) of the photodiode is linearly proportional to the input optical power. Reverse-biased photodetectors remain linear over an extended range (6 decades or more) of photocurrent before saturation occurs. Output saturation occurs at input optical power levels typically greater than 1 milliwatt (mW). Because fiber optic communications systems operate at low optical power levels, detector saturation is generally not a problem.

AVALANCHE PHOTODIODES

An **avalanche photodiode (APD)** is a photodiode that internally amplifies the photocurrent by an avalanche process.

Figure 7.4 shows an example APD structure. In APDs, a large reverse-bias voltage, typically over 100 volts, is applied across the active region. This voltage causes the electrons initially generated by the incident photons to accelerate as they move through the APD active region.

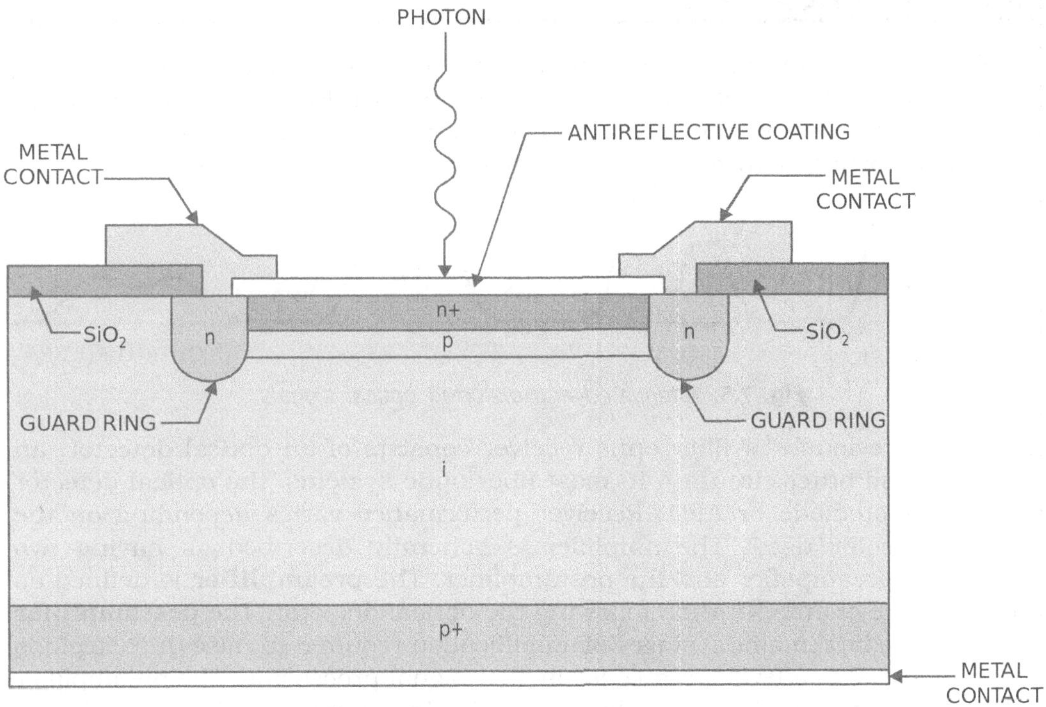

Fig. 7.4. *The basic structure of an APD.*

As these electrons collide with other electrons in the semiconductor material, they cause a fraction of them to become part of the photocurrent. This process is known as **avalanche multiplication.** Avalanche multiplication continues to occur until the electrons move out of the active area of the APD.

The gain of the APD can be changed by changing the reverse-bias voltage. A larger reverse-bias voltage results in a larger gain. However, a larger reverse-bias voltage also results in increased noise levels. Excess noise resulting from the avalanche multiplication process places a limit on the useful gain of the APD. The avalanche process introduces excess noise because every photogenerated carrier does not undergo the same multiplication.

The noise properties of an APD are affected by the materials that the APD is made of. Typical semiconductor materials used in the construction of low-noise APDs include silicon (Si), indium gallium arsenide (InGaAs), and germanium (Ge).

Trade-offs are made in APD design to optimize responsivity and gain, dark current, response time, and linearity. This chapter does not attempt to discuss trade-offs in APD design in more detail. Many aspects of the discussion provided on responsivity, dark current, and response time provided in the PIN photodiodes section also relate to APDs. The response time of an APD and its output circuitry depends on the same factors as PIN photodiodes. The only additional factor affecting the response time of an APD is the additional time required to complete the process of avalanche multiplication.

FIBER OPTIC RECEIVERS

In fiber optic communications systems, optical signals that reach fiber optic receivers are generally attenuated and distorted (see Fig. 7.5). The fiber optic receiver must convert the input and amplify the resulting electrical signal without distorting it to a point that other circuitry cannot use it.

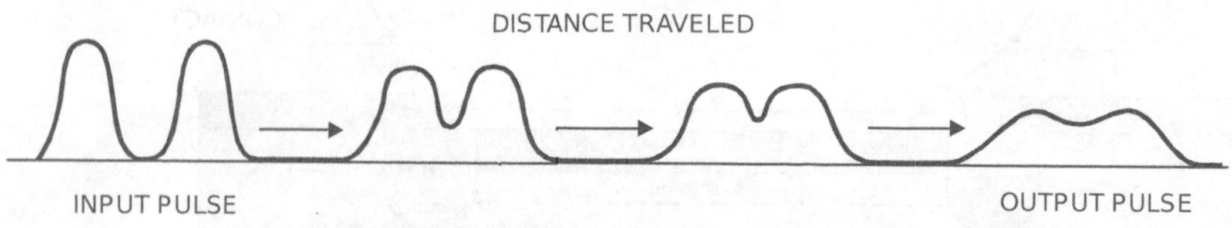

Fig. 7.5. *Attenuated and distorted optical signals.*

As stated previously, a fiber optic receiver consists of an optical detector, an amplifier, and other circuitry. In most fiber optic systems, the optical detector is a PIN photodiode or APD. Receiver performance varies depending on the type of detector used. The amplifier is generally described as having two stages: the preamplifier and the postamplifier. The **preamplifier** is defined as the first stage of amplification following the optical detector. The **postamplifier** is defined as the remaining stages of amplification required to raise the detector's electrical signal to a level suitable for further signal processing. The preamplifier is the dominant contributor of electrical noise in the receiver. Because of this, its design has a significant influence in determining the sensitivity of the receiver.

The output circuitry processes the amplified signal into a form suitable for the interfacing circuitry. For digital receivers, this circuitry may include low-pass filters and comparators. For analog receivers, this circuitry may also include low-pass filters.

Receiver sensitivity, bandwidth, and dynamic range are key operational parameters used to define receiver performance. One goal in designing fiber optic receivers is to optimize receiver sensitivity. To increase sensitivity, receiver noise resulting from signal-dependent shot noise and thermal noise must be kept at a minimum. A more detailed discussion of receiver shot and thermal noise is provided later in this chapter.

In addition to optimizing sensitivity, optical receiver design goals also include optimizing the bandwidth and the dynamic range. A receiver that has the ability to operate over a wide range of optical power levels can operate efficiently in both short- and long-distance applications. Because conflicts arise when attempting to meet each goal, trade-offs in receiver designs are made to optimize overall performance.

RECEIVER NOISE

Noise corrupts the transmitted signal in a fiber optic system. This means that noise sets a lower limit on the amount of optical power required for proper receiver operation. There are many sources of noise in fiber optic systems. They include the following:

- Noise from the light source.
- Noise from the interaction of light with the optical fiber.
- Noise from the receiver itself.

Because the intent of this chapter is to discuss optical detector and receiver properties, only noise associated with the photodetection process is discussed. **Receiver noise** includes thermal noise, dark current noise, and quantum noise. Noise is the main factor that limits receiver sensitivity.

Noise introduced by the receiver is either signal dependent or signal independent. Signal dependent noise results from the random generation of electrons by the incident optical power. Signal independent noise is independent of the incident optical power level.

Thermal noise is the noise resulting from the random motion of electrons in a conducting medium. Thermal noise arises from both the photodetector and the load resistor. Amplifier noise also contributes to thermal noise. A reduction in thermal noise is possible by increasing the value of the load resistor. However, increasing the value of the load resistor to reduce thermal noise reduces the receiver bandwidth. In APDs, the thermal noise is unaffected by the internal carrier multiplication.

Shot noise is noise caused by current fluctuations because of the discrete nature of charge carriers. Dark current and quantum noises are two types of noise that manifest themselves as shot noise. **Dark current noise** results from dark current that continues to flow in the photodiode when there is no incident light. Dark current noise is independent of the optical signal. In addition, the discrete nature of the photodetection process creates a signal dependent shot noise called quantum noise. **Quantum noise** results from the random generation of electrons by the incident optical radiation.

In APDs, the random nature of the avalanche process introduces an additional shot noise called excess noise. For further information on the excess noise resulting from the avalanche process, refer to the avalanche photodiode section.

RECEIVER DESIGN

The simplest fiber optic receivers consist of only the optical detector and a load resistor. However, the output signal of these simple receivers is not in a suitable form for most types of interfacing circuitry. To produce a suitable signal, a preamplifier, a post amplifier, and other circuitry are generally included in the receiver.

The choice of an optical detector and the design of the preamplifier help determine the operational characteristics of the receiver. Fiber optic receivers using APDs have greater sensitivity than those using PIN photodiodes.

In addition, trade-offs are made in preamplifier designs to increase sensitivity while optimizing bandwidth and dynamic range. The two basic types of amplifiers used in fiber optic receivers are the **high-impedance amplifier** and the **transimpedance amplifier.**

The high-impedance preamplifier is generally used with a large load resistor to improve sensitivity. The large load resistor is used to reduce thermal noise. Although the high-impedance preamplifier achieves high sensitivity, receiver bandwidth and dynamic range are limited. The transimpedance preamplifier uses a low-noise, high-input impedance amplifier with negative feedback. This design provides improvements in bandwidth and dynamic range with some degradation in sensitivity from an increase in noise.

FIBER OPTIC RECEIVER PACKAGES

Fiber optic receivers come in packages similar to those for fiber optic transmitters. For information on fiber optic receiver packages, refer back to the fiber optic transmitter packages section of Chapter 6.

FIBER OPTIC RECEIVER APPLICATIONS

Fiber optic receivers can be classified into two categories: **digital** and **analog.** Digital receivers detect the input optical signal, amplify the digital photocurrent, and reshape the signal to produce an undistorted output electrical signal. Analog receivers detect the input optical signal and amplify the generated photocurrent.

Digital Applications

For most digital applications the designs of the digital fiber optic receivers are similar. For **low-data-rate** applications, PIN diodes and high impedance amplifiers are generally used.

Receiver sensitivities are maximized by using large load resistors in the photodiode circuit. For **moderate-data-rate** applications, PIN diodes and either high impedance amplifiers with smaller load resistances or transimpedance amplifiers are used. For **high-data-rate** applications, PINs or APDs are used with transimpedance amplifiers. APDs are rarely used in low- or moderate-data-rate applications unless receivers with extremely low sensitivities are required.

For each digital application, the receiver will generally contain a low-pass filter. The pass-band of the filter depends on the data rate of the application. The filter is used to smooth the amplified signal to remove some of the high frequency noise before the signal is further processed. The digital receiver generally contains a comparator, which reshapes the amplified electrical signal to remove any distortions introduced in the transmission process. In some cases the receiver may also contain clock recovery circuitry, which retimes the output electrical signal as well.

Analog Applications

Analog receivers are similar in design to digital receivers with the exception that digital signal restoring circuitry is not used. The preamplifier and postamplifiers are designed to be more linear than those used in digital receivers in some cases.

For **low-frequency** applications, PIN diodes and high impedance amplifiers are generally used. For **moderate-frequency** applications, PIN diodes and either high impedance amplifiers or transimpedance amplifiers are used. For **high-frequency** applications, PINs or APDs are used with transimpedance amplifiers. As in digital applications, APDs are rarely used in low- or moderate-frequency applications unless receivers with extremely low sensitivities are required.

SUMMARY

Now that you have completed this chapter, let's review some of the new terms, concepts, and ideas that you have learned. You should have a thorough understanding of these principles before moving on to Chapter 8.

A FIBER OPTIC RECEIVER is an electro-optic device that accepts optical signals from an optical fiber and converts them into electrical signals. A typical fiber optic receiver consists of an optical detector, a low-noise amplifier and other circuitry used to produce the output electrical signal.

RECEIVER SPECTRAL RESPONSE, SENSITIVITY, FREQUENCY RESPONSE and DYNAMIC RANGE are key receiver performance parameters that can affect overall system operation.

RECEIVER SENSITIVITY is the minimum amount of optical power required to achieve a specific receiver performance. For digital transmission at a given data rate and coding, this performance is described by a maximum bit-error rate (BER). In analog systems, for a given modulation and bandwidth, it is described by a minimum signal-to-noise ratio (SNR).

DYNAMIC RANGE refers to the range of optical power levels over which the receiver operates within the specified values. It usually is described by the ratio of the maximum input power to the sensitivity.

A TRANSDUCER is a device that converts input energy of one form into output energy of another.

An OPTICAL DETECTOR is a transducer that converts an optical signal into an electrical signal. It does this by generating an electrical current proportional to the intensity of incident optical radiation.

The semiconductor POSITIVE-INTRINSIC-NEGATIVE (PIN) PHOTODIODE and AVALANCHE PHOTODIODE (APD) are the principal optical detectors used in fiber optic systems.

A PHOTOCURRENT is generated when photons are absorbed by a photodiode.

RESPONSIVITY is the ratio of the optical detector's output photocurrent in amperes to the incident optical power in watts.

DARK CURRENT, or reverse-leakage current, is the current that continues to flow in the photodetector when there is no incident light.

The RESPONSE TIME of a photodiode and its output circuitry depends primarily on the thickness of the detector active area and the detector RC time constant.

The TRANSIT TIME is the time it takes electrons to travel out of the detector active area.

The RC TIME CONSTANT is defined by the capacitance (C) of the photodiode and the resistance (R) of the load. The RC time constant is given by $t_{RC} = RC$.

A HIGH-SPEED RESPONSE requires short transit times and low capacitance. However, any change in photodiode parameters to optimize the transit time and capacitance can also affect quantum efficiency, dark current, and coupling efficiency.

Detector LINEARITY means that the output electrical current (photocurrent) of the photodiode is linearly proportional to the input optical power.

An AVALANCHE PHOTODIODE (APD) is a photodiode that internally amplifies the photocurrent by an avalanche process.

In APDs, a large REVERSE-BIAS VOLTAGE, typically over 100 volts, is applied across the active region.

AVALANCHE MULTIPLICATION occurs when accelerated electrons collide with other electrons in the semiconductor material, causing a fraction of them to become part of the photocurrent.

TRADE-OFFS are made in APD design to optimize responsivity and gain, dark current, response time and linearity.

The RESPONSE TIME of APDs accounts for the avalanche build-up time in addition to transit time and RC time constant.

The PREAMPLIFIER is defined as the first stage of amplification following the optical detector.

The POSTAMPLIFIER is defined as the remaining stages of amplification required to raise the detectors electrical signal to a level suitable for further signal processing.

RECEIVER SENSITIVITY, BANDWIDTH and DYNAMIC RANGE are key operational parameters used to define receiver performance.

NOISE is the main factor that determines receiver sensitivity.

RECEIVER NOISE includes thermal noise, dark current noise and quantum noise.

THERMAL NOISE is the noise resulting from the random motion of electrons in a conducting medium.

SHOT NOISE is noise caused by current fluctuations due to the discrete nature of charge carriers.

DARK CURRENT NOISE results from dark current that continues to flow in the photodiode when there is no incident light.

QUANTUM NOISE results from the random generation of electrons by the incident optical radiation.

The HIGH-IMPEDANCE AMPLIFIER and the TRANSIMPEDANCE AMPLIFIER are the two basic types of amplifiers used in fiber optic receivers.

The HIGH-IMPEDANCE PREAMPLIFIER provides a high sensitivity, but limits receiver bandwidth and dynamic range.

The TRANSIMPEDANCE PREAMPLIFIER provides improvements in bandwidth and dynamic range with some degradation in sensitivity from an increase in noise.

PIN PHOTODIODES are used as the detector in most applications.

AVALANCHE PHOTODIODES are only used in high-speed applications and applications where detectors with extremely low sensitivities are required.

REVIEW QUESTIONS AND ANSWERS

Q. 1. *What is a fiber optic receiver?*

Ans. An electro-optic device that accepts optical signals from an optical fiber and converts them into electrical signals.

Q. 2. *Which part of the receiver amplifies the electrical signal to a level suitable for further signal processing?*

Ans. Amplifier.

Q. 3. *Which performance parameter is the minimum amount of optical power required to achieve a specific bit-error rate (BER) in digital systems or a given signal-to-noise ratio (SNR) in analog systems?*

Ans. Receiver sensitivity.

Q. 4. *Define receiver dynamic range.*

Ans. The range of optical power levels over which the receiver operates within the specified values. It is usually described by the ratio of the maximum input power to the sensitivity.

Q. 5. *Describe the operation of an optical detector.*

Ans. It is a transducer that converts an optical signal into an electrical signal. It does this by generating an electrical current proportional to the intensity of incident optical radiation.

Q. 6. *For efficient operation, should a detector have a high or low responsivity at the operating wavelength?*

Ans. High.

Q. 7. *List the two principal optical detectors used in fiber optic systems.*

Ans. The semiconductor positive-intrinsic-negative (PIN) photodiode and avalanche photodiode (APD).

Q. 8. *What are the four most common materials used in semiconductor detector fabrication?*

Ans. Silicon, gallium arsenide, germanium and indium phosphide.

Q. 9. *What is a photocurrent?*

Ans. The current produced when photons are incident on the detector active area.

Q. 10. *Define responsivity.*

Ans. The ratio of the optical detector's output photocurrent in amperes to the incident optical power in watts.

Q. 11. *How are PIN photodiodes usually biased?*

Ans. Reverse-biased.

Q. 12. *What is the dark current?*

Ans. The leakage current that continues to flow through a photodetector when there is no incident light.

Q. 13. *Will dark current increase or decrease as the temperature of the photodiode increases?*

Ans. Increase.

Q. 14. *Should the capacitance of the photodetector be kept small or large to prevent the RC time constant from limiting the response time?*

Ans. Small.

Q. 15. *Trade-offs between competing effects are necessary for high speed response. Which competing effect (fast transit time, low capacitance, or high quantum efficiency) requires a thin active area?*

Ans. Fast transit time.

Q. 16. *Why is detector saturation not generally a problem in fiber optic communications systems?*

Ans. Because fiber optic communications systems operate at low optical power levels.

Q. 17. *Describe avalanche multiplication.*

Ans. The electrons initially generated by the incident photons accelerate as they move through the APD active region. As these electrons collide with electrons in the semiconductor material, they cause a fraction of them to become part of the photocurrent.

Q. 18. *How can the gain of an APD be increased?*

Ans. By increasing the reverse-bias voltage.

Q. 19. *Which amplifier stage (the preamplifier or the postamplifier) is a dominant contributor of noise and significantly influences the sensitivity of the receiver?*

Ans. The preamplifier.

Q. 20. *List the key operational parameters used to define receiver performance.*

Ans. Receiver sensitivity, bandwidth and dynamic range.

Q. 21. *List the main types of receiver noise.*

Ans. Thermal noise, dark current noise and quantum noise.

Q. 22. *What is the main factor that determines receiver sensitivity?*

Ans. Noise.

Q. 23. *For a reduction in thermal noise, should the value of the detector's load resistor be increased or decreased?*

Ans. Increased.

Q. 24. *What are two types of noise that manifest themselves as shot noise?*

Ans. Dark current and quantum noises.

Q. 25. *What are the two basic types of preamplifiers used in fiber optic receivers?*

Ans. The high-impedance amplifier and the transimpedance amplifier.

Q. 26. *Which preamplifier design (high-impedance or transimpedance) provides improvements in bandwidth and greater dynamic range with some degradation in sensitivity from an increase in noise?*

Ans. Transimpedance.

Q. 27. *For what types of applications are APDs generally used?*

Ans. For high-data-rate applications and for low- or moderate-data-rate applications where receivers with extremely low sensitivities are required.

Q. 28. *Why is a low-pass filter generally part of a digital fiber optic receiver?*

Ans. To smooth the amplified signal to remove some of the high frequency noise before the signal is further processed.

■■■

8 FIBER OPTIC LINKS

Learning Objectives

Upon completion of this chapter, you should be able to do the following:

- Describe a basic point-to-point fiber optic data link.
- Explain the difference between digital and analog fiber optic communications systems.
- Discuss the most common types of line coding used in digital fiber optic communications including non-return-to-zero (NRZ), return-to-zero (RZ), and biphase (or Manchester).
- Describe the main type of analog modulation.
- State several precautions that need to be emphasized when installing fiber optic links on board ships.

FIBER OPTIC SYSTEM TOPOLOGY

Most of the discussion on fiber optic data links provided earlier in this training manual refers to simple point-to-point links. A **point-to-point** fiber optic data link consists of an optical transmitter, optical fiber, and an optical receiver. In addition, any splices or connectors used to join individual optical fiber sections to each other and to the transmitter and the receiver are included. Figure 8.1 provides a schematic diagram of a point-to-point fiber optic data link.

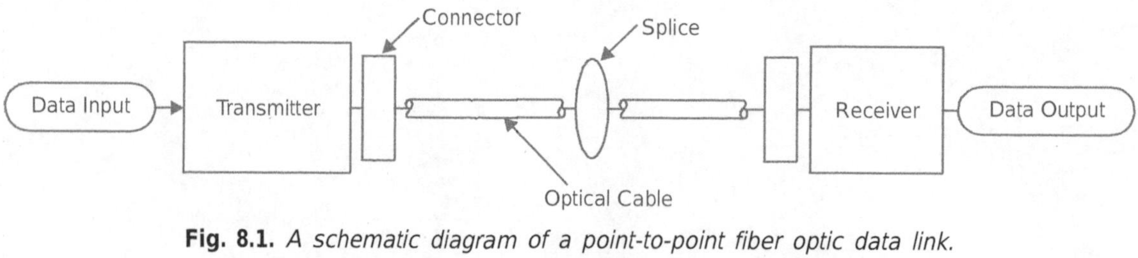

Fig. 8.1. *A schematic diagram of a point-to-point fiber optic data link.*

A common fiber optic application is the **full duplex link.** This link consists of two simple point-to-point links. The links transmit in opposite directions between the equipments. This application may be configured using only one fiber. If configured with one fiber, fiber optic splitters are used at each end to couple the transmit signal onto the fiber and receive signal to the detector.

All fiber optic systems are simply sets of point-to-point fiber optic links. Different system topologies arise from the different ways that point-to-point fiber optic links can be connected between equipments. The term **topology,** as used here, refers to the configuration of various equipments and the fiber optic components interconnecting them. This equipment may be computers, workstations, consoles, or other equipments. Point-to-point links are connected to produce systems with linear bus, ring, star, or tree topologies. Point-to-point fiber optic links are the basic building block of all fiber optic systems.

A **linear bus topology** consists of a single transmission line that is shared by a number of equipments (see Fig. 8.2). Generally the transmission line in a fiber optic linear bus consists of two optical lines, one for each direction of communication.

Optical taps (optical splitters) are used by each equipment to connect to each line. For each line, the optical tap couples signals from the line to the equipment receiver and from the equipment transmitter onto the line. The connection between any two equipments is a simple point-to-point link that contains the optical tap for each equipment.

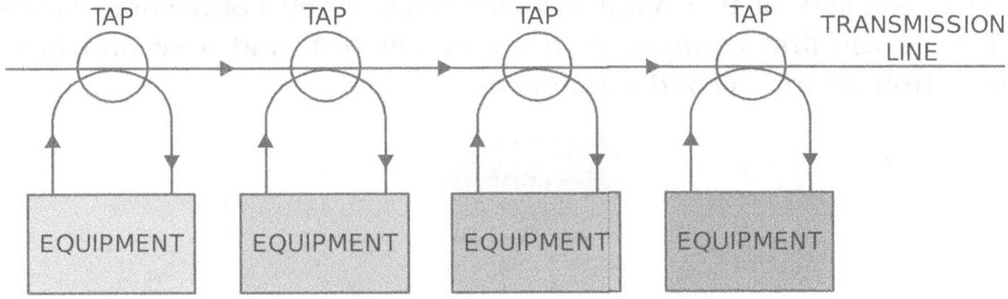

Fig. 8.2. *Linear bus topology.*

A **ring topology** consists of equipments attached to one another in a closed loop or ring (see Fig. 8.3).

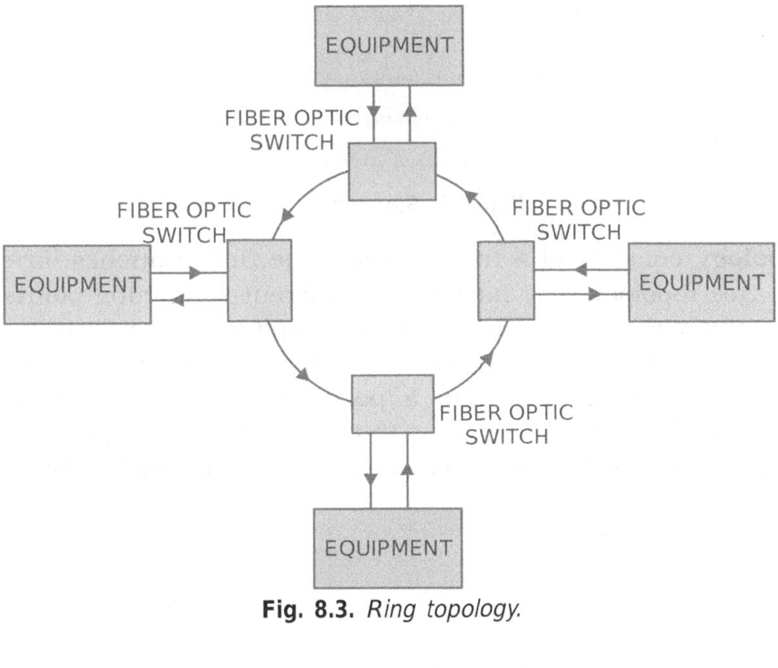

Fig. 8.3. *Ring topology.*

The connection between each equipment is a simple point-to-point link. In some systems each equipment may have an associated optical switch. In normal operation, the switch routes signals from the fiber connected to the previous equipment to the receiver. It also routes signals from the transmitter to the fiber connected to the next equipment. In bypass operation, the switch routes signals from the fiber connected to the previous equipment to the fiber connected to the next equipment. In each case, the connection between adjacent equipments on the ring is a simple point-to-point link through fiber, connectors, and switches.

In the **star topology,** each equipment is connected to a common center hub (see Fig. 8.4). The center hub can be a passive fiber optic star coupler or an active equipment. If the center hub is a passive star coupler, each equipment transmitter is connected to an input port of the coupler and an output port of the coupler is connected to each equipment receiver. The connection between any two equipments is a simple point-to-point link through the star coupler. If the center hub is an active equipment, the connection between any two equipments consists of two point-to-point links. Each connection consists of one link from the first equipment to the center hub and a second link from the center hub to the second equipment.

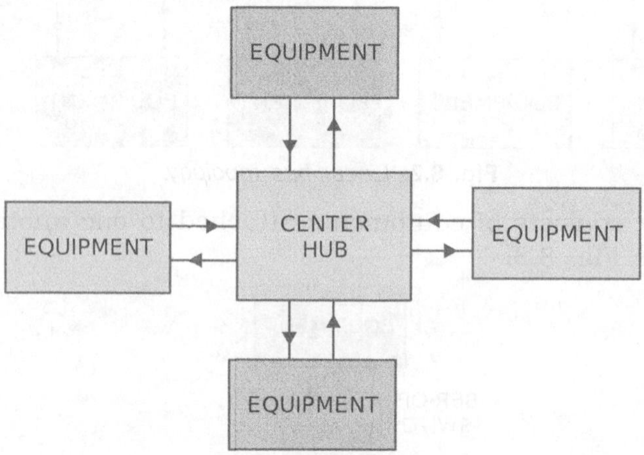

Fig. 8.4. *Star topology.*

A tree topology consists of a transmission line that branches, or splits (see Fig. 8.5). A tree topology may have many different branching points. At each branching point either a passive fiber optic splitter or an active branching device is used. In many cases both passive couplers and active branching devices are used within a particular system. Regardless of the branching method, each connection within the tree is a simple point-to-point link through splitters or multiple point-to-point links through active branching devices.

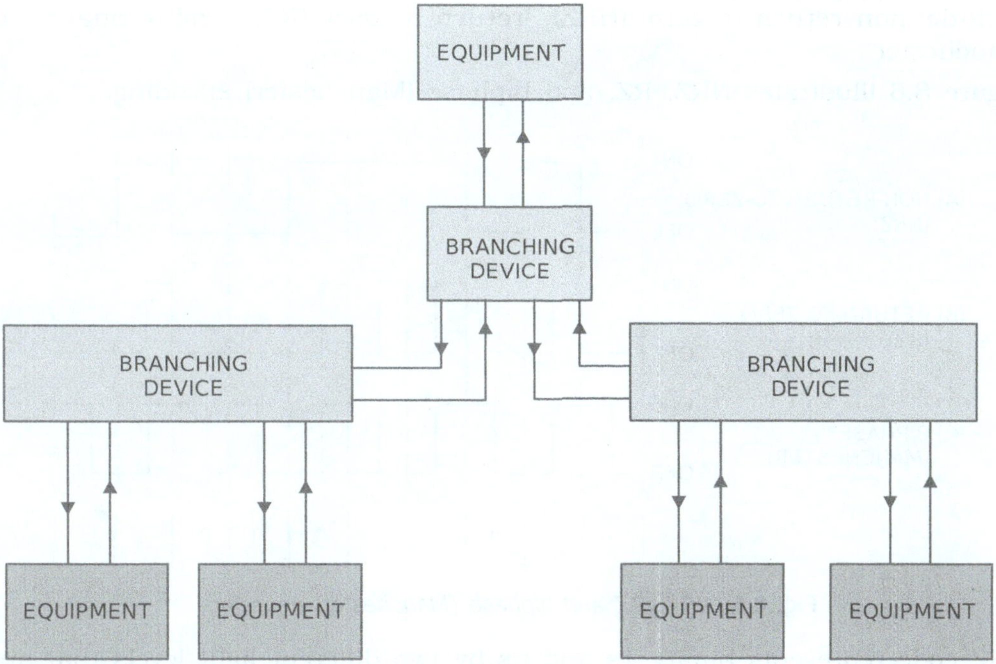

Fig. 8.5. *Tree topology.*

LINK CLASSIFICATION

While there are several ways to classify fiber optic links, this chapter classifies links according to the modulation type: either **digital** or **analog. Modulation** is the process of varying one or more characteristics of an optical signal to encode and convey information. Generally, the intensity of the optical signal is modulated in fiber optic communications systems.

- Digital modulation implies that the optical signal consists of discrete levels.
- Analog modulation implies that the intensity of the optical signal is proportional to a continuously varying electrical input. Most fiber optic systems are digital because digital transmission systems generally provide superior performance over analog transmission systems.

DIGITAL TRANSMISSION

A **digital signal** is a discontinuous signal that changes from one state to another in discrete steps. A popular form of digital modulation is **binary** or two level, digital modulation. In binary modulation the optical signal is switched from a low-power level (usually off) to a high-power level. There are a number of modulation techniques used in digital systems, but these will not be discussed here.

Line coding is the process of arranging symbols that represent binary data in a particular pattern for transmission.

The most common types of line coding used in fiber optic communications include non-return-to-zero (NRZ), return-to-zero (RZ), and biphase, or Manchester.

Figure 8.6 illustrates NRZ, RZ, and biphase (Manchester) encoding.

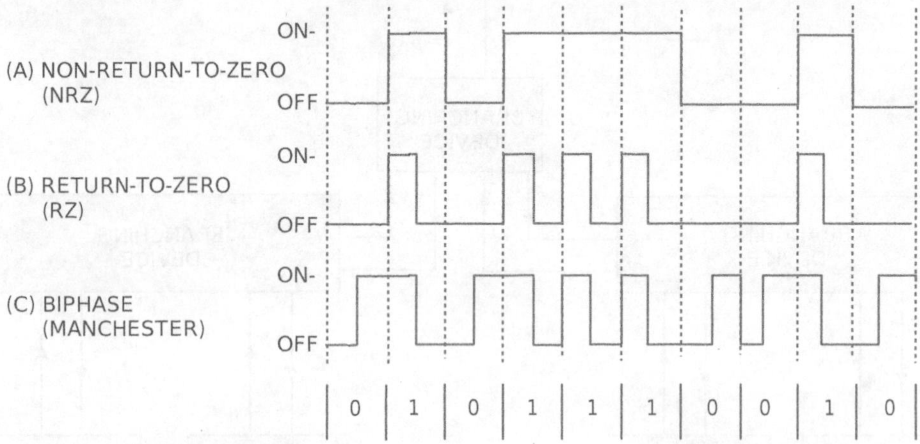

Fig. 8.6. *NRZ, RZ and biphase (Manchester) encoding.*

NRZ code represents binary 1s and 0s by two different light levels that are constant during a bit duration. The presence of a high-light level in the bit duration represents a binary 1, while a low-light level represents a binary 0. NRZ codes make the most efficient use of system bandwidth. However, loss of timing may result if long strings of 1s and 0s are present causing a lack of level transitions. RZ coding uses only half the bit duration for data transmission.

In RZ encoding, a half period optical pulse present in the first half of the bit duration represents a binary 1. While an optical pulse is present in the first half of the bit duration, the light level returns to zero during the second half. A binary 0 is represented by the absence of an optical pulse during the entire bit duration. Because RZ coding uses only half the bit duration for data transmission, it requires twice the bandwidth of NRZ coding. Loss of timing can occur if long strings of 0s are present.

Biphase or Manchester, encoding incorporates a transition into each bit period to maintain timing information. In Manchester encoding, a high-to-low light level transition occurring in the middle of the bit duration represents a binary 1. A low-to-high light level transition occurring in the middle of the bit duration represents a binary 0.

Digital transmission offers an advantage with regard to the acceptable signal-to-noise ratio (SNR) at the optical receiver. Digital communications systems can tolerate large amounts of signal loss and dispersion without impairing the ability of the receiver to distinguish a binary 1 from a binary 0. Digital signalling also reduces the effects that optical source nonlinearities and temperature have on system performance. Source nonlinearities and temperature variations can severely affect analog transmission. Digital transmission provides superior performance in most complex systems (such as LANs) and long-haul communications systems. In short-haul systems, the cost and complexity of analog-to-digital and digital-to-analog conversion equipment, in some cases, outweigh the benefits of digital transmission.

ANALOG TRANSMISSION

An **analog signal** is a continuous signal whose amplitude, phase, or some other property varies in a direct proportion to the instantaneous value of a physical variable. An example of an analog signal is the output power of an optical source whose intensity is a function of a continuous electrical input signal.

Most analog fiber optic communications systems intensity modulate the optical source. In **intensity modulation,** the intensity of the optical source's output signal is directly modulated by the incoming electrical analog baseband signal. A **baseband signal** is a signal that is in its original form and has not been changed by a modulation technique.

In some cases, the optical source may be directly modulated by a incoming electrical signal that is not a baseband signal. In these cases the original electrical signal generally modulates an electrical subcarrier frequency. The most common form of analog subcarrier modulation in fiber optic systems is frequency modulation (FM). The optical source is intensity modulated by the electrical subcarrier.

While most fiber optic systems employ digital modulation techniques, there are certain applications where analog modulation techniques are preferred.

The transmission of video using analog techniques is very popular, especially for shorter distances, where costs can be minimized and complex multiplexing and timing equipment is unnecessary. The transmission of analog voice signals may also be attractive in small, short-haul systems. In addition, fiber optic sensor systems may incorporate analog transmission. Requirements that analog transmission places on applications include high signal-to-noise ratio and high source linearity. While analog transmission can be attractive for short-haul or medium-haul systems, it is unattractive for long-haul systems where digital techniques provide better performance.

SYSTEM DESIGN

Fiber optic systems can be simple point-to-point data links or can involve more complex topologies. However, it is generally necessary only to refer to point-to-point data links when discussing the process of link design. Fiber optic systems that incorporate complex architectures can be simplified into a collection of point-to-point data links before beginning the design process.

Fiber optic system design is a complicated process that involves link definition and analysis. The design process begins by providing a complete description of the communication requirements. This information is used to develop the link architecture and define the communications links. System designers must decide on the operational wavelength and types of components to use in the system. These decisions affect numerous system and link design parameters, such as launched power, connection losses, bandwidth, cost and reliability.

Once a system design has been formulated, each link is analyzed to determine its viability. Link analysis involves calculating each link's power budget and risetime budget. Calculating a **power budget** involves identifying all of the sources of loss in the fiber optic link. These losses and an additional safety margin are then compared to the difference between the transmitter output power and the receiver sensitivity. The difference between the transmitter output power and the receiver sensitivity is referred to as the **available power.** If the sources of loss plus the safety margin are less than the available power in the link, the design is viable.

Calculating the **risetime budget** involves calculating the risetimes of the link transmitter and the optical fiber. The composite optical transmitter/fiber risetime is referred to as the **fiber exit risetime.** If the fiber exit risetime is less than the maximum input risetime specified for the link receiver, then the link design is viable.

If a proposed link design is not viable, the system designer will reevaluate various decisions made earlier in the system design. These reevaluations may include using a different transmitter or receiver or may involve redesigning the physical configuration of the link. Because there are many variables involved in link design and analysis, it may take several iterations before the variables are combined in a manner that ensures link operation. For more information of fiber optic system design, refer to the *Navy Fiber Optic* and *System Design Standard.*

SYSTEM INSTALLATION

The Navy has a standard to provide detailed information and guidance to personnel concerned with the installation of fiber optic cable plants on naval surface ships and submarines. The **fiber optic cable plant** consists of all the fiber optic cables and the fiber optic interconnection equipment within the ship, including connectors, splices and interconnection boxes. The fiber optic cable plant installation standard consists of a basic standard and six numbered parts dealing with the following:

- Cables—provides detailed methods for cable storage and handling, end-sealing, repair, and splicing.
- Equipment—provides detailed methods for fiber optic equipment installation and cable entrance to equipment.
- Penetrations—provides detailed methods for cable penetrations within the ship's structure.

- Cableways—provides detailed methods to install fiber optic cables in cableways.
- Connectors and interconnections—provides detailed methods for installing fiber optic connectors and other interconnections, such as splices.
- Tests—identifies and provides detailed methods for testing fiber optic cable plants before, during, and after installation and repair.

There are other standards that discuss fiber optic system installation.

Many of these standards incorporate procedures for repair, maintenance and testing. The techniques developed for installing fiber optic hardware are not much different than for installing hardware for copper-based systems.

However, the primary precautions that need to be emphasized when installing fiber optic systems on board ships are as follows:

- Optical fibers or cables should never be bent at a radius of curvature less than a certain value, called the **minimum bend radius.** Bending an optical fiber or cable at a radius smaller than the minimum bend radius causes additional fiber loss.
- Fiber optic cables should never be pulled tight or fastened over or through sharp corners or cutting edges. Extremely sharp bends increase the fiber loss and may lead to fiber breakage.
- Fiber optic connectors should always be cleaned before mating. Dirt in a fiber optic connection will significantly increase the connection loss and may damage the connector.
- Precautions must be taken so the cable does not become kinked or crushed during installation of the hardware. Extremely sharp kinks or bends increase the fiber loss and may lead to fiber breakage.
- Only trained, authorized personnel should be allowed to install or repair fiber optic systems.

SUMMARY

Now that you have completed this chapter, let's review some of the new terms, concepts and ideas that you have learned. Understanding the basics of fiber optic system classification, design and installation is recommended before you begin studying specific fiber optic system applications.

A basic POINT-TO-POINT fiber optic data link consists of an optical transmitter, optical fiber and an optical receiver.

In addition, any splices or connectors used to join individual optical fiber sections to each other and to the transmitter and the receiver are included.

The term TOPOLOGY refers to the configuration of various equipments and the fiber optic components interconnecting them.

A LINEAR BUS TOPOLOGY consists of a single transmission line that is shared by a number of equipments.

A RING TOPOLOGY consists of equipments attached to one another in a closed loop or ring.

In the STAR TOPOLOGY, each equipment is connected to a common center hub. The center hub can be a passive fiber optic star coupler or an active equipment.

A TREE TOPOLOGY consists of a transmission line that branches, or splits.

FIBER OPTIC LINKS are classified according to the modulation type: either digital or analog.

DIGITAL MODULATION implies that the optical signal consists of discrete levels.

ANALOG MODULATION implies that the intensity of the optical signal is proportional to a continuously varying electrical input.

MODULATION is the process of varying one or more characteristics of an optical signal to encode and convey information.

A DIGITAL SIGNAL is a discontinuous signal that changes from one state to another in discrete steps.

BINARY or two level, digital modulation is a popular form of digital modulation.

REVIEW QUESTIONS AND ANSWERS

Q. 1. *List four system topologies that can be constructed using point-to-point fiber optic links.*

Ans. Linear bus, ring, star, and tree topologies.

Q. 2. *Which topology (linear bus, ring, star, or tree) consists of equipments attached to one another in a closed loop?*

Ans. Ring.

Q. 3. *Which topology (bus, ring, star, or tree) has a center hub interconnecting the equipments ?*

Ans. Star.

Q. 4. *Define modulation.*

Ans. The process of varying one or more characteristics of an optical signal to encode and convey information.

Q. 5. *What is a digital signal?*

Ans. A discontinuous signal that changes from one state to another in discrete steps.

Q. 6. *In NRZ code, does the presence of a high-light level in the bit duration represent a binary 1 or a binary 0?*

Ans. Binary 1.

Q. 7. *How can the loss of timing occur in NRZ line coding?*

Ans. If long strings of 1s or 0s are present causing a lack of level transitions.

Q. 8. *How is a binary 1 encoded in RZ line coding?*

Ans. A half-period optical pulse present in the first half of the bit duration.

Q. 9. *In Manchester encoding, does a low-to-high light level transition occurring in the middle of the bit duration represent a binary 1 or a binary 0?*

Ans. Binary 0.

Q. 10. *What is an analog signal?*

Ans. A continuous signal that varies in a direct proportion to the instantaneous value of a physical variable.

Q. 11. *What type of modulation do most analog fiber optic communications systems use?*

Ans. Intensity modulation.

Q. 12. *Why has the transmission of video using analog techniques been very popular, especially for shorter distances?*

Ans. Because cost can be minimized and complex multiplexing and timing equipment is unnecessary.

Q. 13. *Why is it generally only necessary to refer to point-to-point data links when discussing the process of fiber optic system design?*

Ans. Because fiber optic systems that incorporate complex architectures can be simplified into a collection of point-to-point data links before beginning the design process.

Q. 14. *List five system design parameters considered when system designers choose the system operational wavelength and link components.*

Ans. Launch power, connection losses, bandwidth, cost and reliability.

Q. 15. *What two analyses are performed to determine if a link design is viable?*

Ans. Power budget and risetime budget.

Q. 16. *Optical fibers or cables should never be bent at a radius of curvature smaller than a certain value. Identify this radius of curvature.*

Ans. Minimum bend radius.

Q. 17. *List five precautions to take when installing fiber optic systems on board naval ships.*

Ans.
- Never bend an optical fiber or cable at a radius of curvature less than the minimum bend radius.
- Never pull fiber optic cables tight or fasten them over or through sharp corners or cutting edges.
- Always clean fiber optic connectors before mating.
- Do not kink or crush fiber optic cable during installation of the hardware.
- Allow only trained, authorized personnel to install or repair fiber optic systems.

INDEX

A

B

C

D

E

F

G

H

I

L

M

N

O

P

Q

R

S

T

U

V

X

Y

❑❑❑